200 KAKURO

PUZZLE BOOK FOR ADULTS

PUZZLES WITH SOLUTIONS
& STARTER NUMBERS

How To Solve Kakuro Puzzles?

Introduction:

A KAKURO puzzle is just like a standard crossword, but the digits 1 to 9 are used instead of the letters of the alphabet. Note that the number "0" cannot be used!

Your mission is to insert a digit from 1 to 9 inclusive into each white cell so that the sum of the numbers in each entry matches the clue associated with it. A helpful key rule is that the same number cannot be repeated within any run, so the solution to the value '4' cannot be 2+2 but must 1+3 (or 3+1).

Useful to solve Kakuro puzzles:

1 - Unique Blocks:

There are some unique definitions that can be solved only in a specific way, and you can call them Unique Blocks for example, if 6 is the sum-clue of a block of three squares then the block must consist of the numbers 1+2+3 but not necessarily in this order.

When spotting a Unique block, the list in the next page can be useful to identify which numbers must be used in that block. The only thing left is to find out in which order the numbers should be organized.

Here is a list of some Unique Blocks:

across two cells:

the sum 3 = 1 + 2

the sum 4 = 1 + 3

the sum 16 = 7 + 9

the sum 17 = 8 + 9

across three cells:

the sum 6 = 1 + 2 + 3

the sum 7 = 1 + 2 + 4

the sum 23 = 6 + 8 + 8

the sum 24 = 7 + 8 + 9

across four cells:

the sum 10 = 1 + 2 + 3 + 4

the sum 11 = 1 + 2 + 3 + 5

the sum 29 = 5 + 7 + 8 + 9

the sum 30 = 6 + 7 + 8 + 9

2- Are there any combinations that intersect and share a unique digit?

Yes, for example, when 5 in two cells crosses 21 in three cells, the common digit must be 4.

For example, you can make 5 in two by using (1,4) or (2,3), and you can make 21 in three by using (9,7,5) or (9,8,4); but the only digit these sets have in common is 4.

Using this tip, you can always analyze intersected clues and study them by looking at the sets of possible digits that make up each sum, then by studying which two sets have a single digit in common.

Conclusion

Practicing is the best way to see how these advices can be applied in a Kakuro puzzle. We wish you good luck and lots of fun!

The table in the next few pages will give you all the kakuro blocks between 2 and 9 cells in length

All Sets

No of Cells	Sum	Sets		Always Used	Never Used
2	3	12		12......	.3456789
2	4	13		1.3.....	.2456789
2	5	14 23		....	...56789
2	6	15 24		....	.3.6789
2	7	16 25 34		....	789
2	8	17 26 35		....	...4..89
2	9	18 27 36 45		....	9
2	10	19 28 37 46		....	...5...
2	11	29 38 47 56		....	1.......
2	12	39 48 57		....	12..6...
2	13	49 58 67		....	123....
2	14	59 68		....	1234..7..
2	15	69 78		....	12345....
2	16	79		....7.9	123456.8.
2	17	89		89	1234567..
3	6	123		123.....	...456789
3	7	124		12.4....	.3.56789
3	8	125 134		1.......	...6789
3	9	126 135 234		....	789
3	10	127 136 145 235		....	89
3	11	128 137 146 236 245		....	9
3	12	129 138 147 156 237 246 345		....	....
3	13	139 148 157 238 247 256 346		....	....
3	14	149 158 167 239 248 257 347 356		....	....
3	15	159 168 249 258 267 348 357 456		....	....

3	16	169 178 259 268 349 358 367 457		.....	.....
3	17	179 269 278 359 368 458 467		......	......
3	18	189 279 369 378 459 468 567		......	......
3	19	289 379 469 478 568		......	1.......
3	20	389 479 569 578		......	12......
3	21	489 579 678		9	123....
3	22	589 679		...6.89	1234....
3	23	689			12345.7..
3	24	789		789	123456...
4	10	1234		1234.....	...56789
4	11	1235		123.5....	...4.6789
4	12	1236 1245		12......	789
4	13	1237 1246 1345		1.......	89
4	14	1238 1247 1256 1346 2345		......	9
4	15	1239 1248 1257 1347 1356 2346		......	......
4	16	1249 1258 1267 1348 1357 1456 2347 2356		......	......
4	17	1259 1268 1349 1358 1367 1457 2348 2357 2456		......	......
4	18	1269 1278 1359 1368 1458 1467 2349 2358 2367 2457 3456		......	......
4	19	1279 1369 1378 1459 1468 1567 2359 2368 2458 2467 3457		......	......
4	20	1289 1379 1459 1478 1568 2369 2378 2459 2468 2567 3458 3467		......	......
4	21	1389 1479 1569 1578 2379 2469 2478 2568 3459 3468 3567		......	......
4	22	1489 1579 1678 2389 2479 2569 2578 3469 3478 3568 4567		......	......
4	23	1589 1679 2439 2579 2678 3479 3569 3578 4568		......	......
4	24	1689 2589 2679 3489 3579 3678 4569 4578		......	......
4	25	1789 2689 3589 3679 4579 4678		......	......
4	26	2789 3689 4539 4679 5678		......	1.......
4	27	3789 4689 5679		9	12......
4	28	4789 5689		89	123......

n	k	Combinations	Pattern A	Pattern B
4	29	5789	...5.789	1234.6...
4	30	6789	6789	12345....
		(shaded)		
5	15	12345	12345....	6789
5	16	12346	1234.6...	...5.789
5	17	12347 12356	123......	89
5	18	12348 12357 12456	12......	9
5	19	12349 12358 12367 12457 13456	1.......	....
5	20	12359 12368 12458 12467 13457 23456	....	....
5	21	12369 12378 12459 12468 12567 13458 13467 23457	....	....
5	22	12379 12469 12478 12568 13459 13468 13567 23458 23467	....	....
5	23	12389 12479 12569 12578 13469 13478 13568 14567 23459 23468 23567	....	....
5	24	12489 12579 12678 13479 13569 13578 14568 23469 23478 23568 24567	....	....
5	25	12589 12679 13489 13579 13678 14569 14578 23489 23579 23678 24569 24578 34567	....	....
5	26	12689 13589 13679 14589 14679 15678 23689 24589 24679 25678 34579	....	....
5	27	12789 13689 14589 14679 15678 23589 23679 24579 24678 34569 34578	....	....
5	28	13789 14689 15679 23689 24589 24679 25678 34579 34678	....	....
5	29	14789 15689 23789 24689 25679 34589 34679 35678	....	....
5	30	15789 24789 25689 34689 35679 45678	....	9
5	31	16789 25789 34789 35689 45679	9	89
5	32	26789 35789 45689	89	789
5	33	36789 45789	789	..4.6789
5	34	46789	..4.6789	...56789
5	35	56789	...56789	
		(shaded)		
6	21	123456	123456...	789
6	22	123457	12345.7...	...6.89
6	23	123458 123467	1234.....	9
6	24	123459 123468 123567	123......	....

n	#	Combinations		
6	25	123469 123478 123568 124567	12......	......
6	26	123479 123569 123578 124568 134567	1......	......
6	27	123489 123579 123678 124569 124578 134568 234567	......	......
6	28	123589 123679 124579 124678 134569 134578 234568	......	......
6	29	123689 124589 124679 125678 134579 134678 234569 234578	......	......
6	30	123789 124689 125679 134589 134679 135678 234579 234678	......	......
6	31	124789 125689 134689 135679 145678 234589 234679 235678	......	......
6	32	125789 134789 135689 145679 234689 235679 245678	9	......
6	33	126789 135789 145689 234789 235689 245679 345678	89	......
6	34	136789 145789 235789 245689 345679	789	9
6	35	146789 236789 245789 345689	6789	89
6	36	156789 246789 345789	...6789	789
6	37	256789 346789	......	6789
6	38	356789	..3.56789	......
6	39	456789	...456789	......
7	28	1234567	1234567..	89
7	29	1234568	123456.8.	....7.9
7	30	1234569 1234578	12345....	......
7	31	1234579 1234678	1234..7..	......
7	32	1234589 1234679 1235678	123......	......
7	33	1234689 1235679 1245678	12..6....	......
7	34	1234789 1235689 1245679 1345678	1......	......
7	35	1235789 1245689 1345679 2345678	...5....	......
7	36	1236789 1245789 1345689 2345679	9	......
7	37	1246789 1345789 2345689	..4..89	......
7	38	1256789 1346789 2345789	789	......
7	39	1356789 2346789	..3.6789	......
7	40	1456789 2356789	...56789	......

7	41	2456789		.2.456789	1.3.....
7	42	3456789		..3456789	12......
8	36	12345678		12345678	9
8	37	12345679		12345567.9	.....8.
8	38	12345689		123456.89	...7.
8	39	12345789		12345.789	...6...
8	40	12346789		1234.6789	...5...
8	41	12356789		123.56789	.4.....
8	42	12456789		12.456789	.3.....
8	43	13456789		1.3456789	.2.....
8	44	23456789		23456789	1.......
9	45	123456789		123456789	....

KAKURO

PUZZLE BOOK FOR ADULTS

-EASY-

5X5

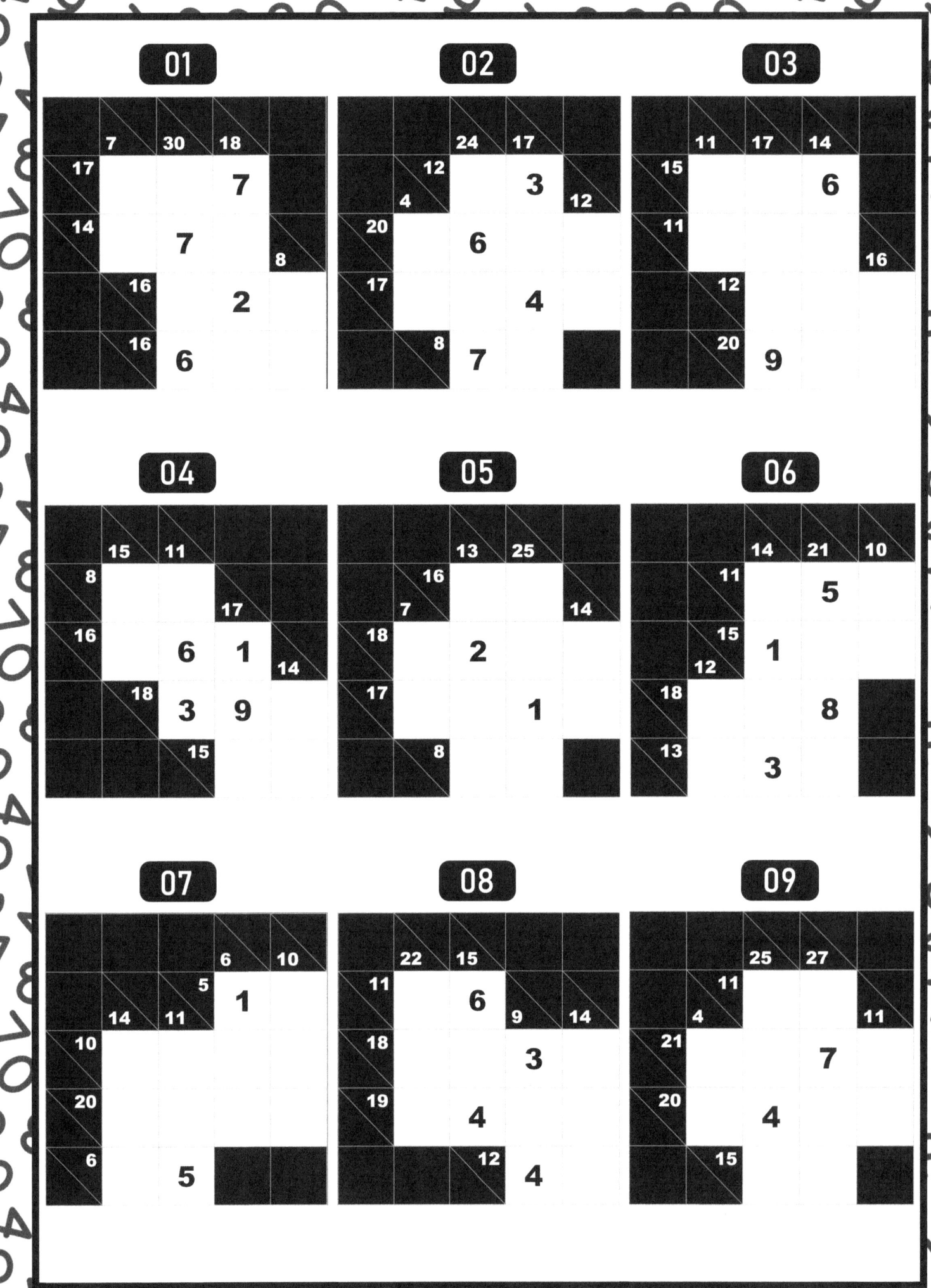

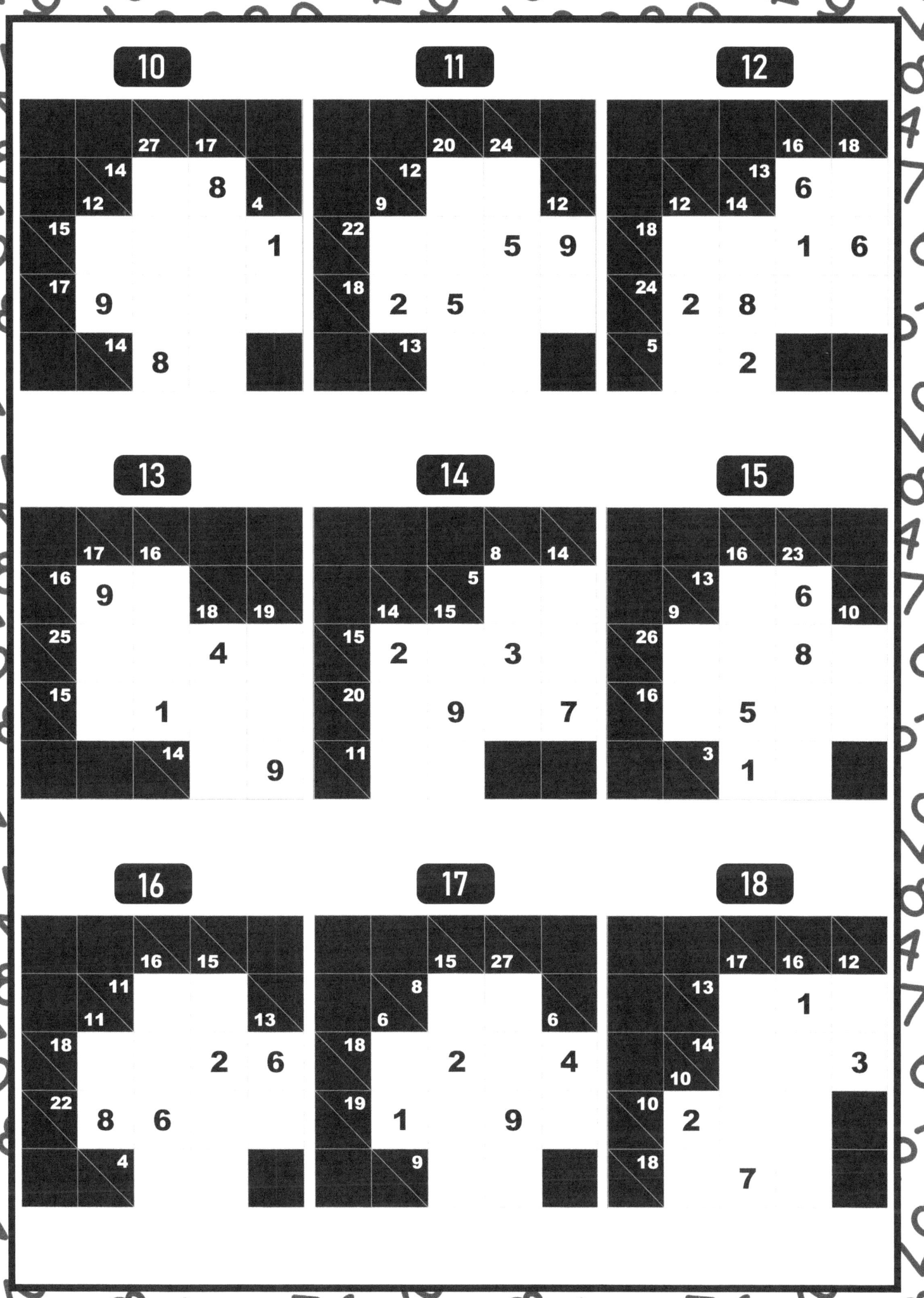

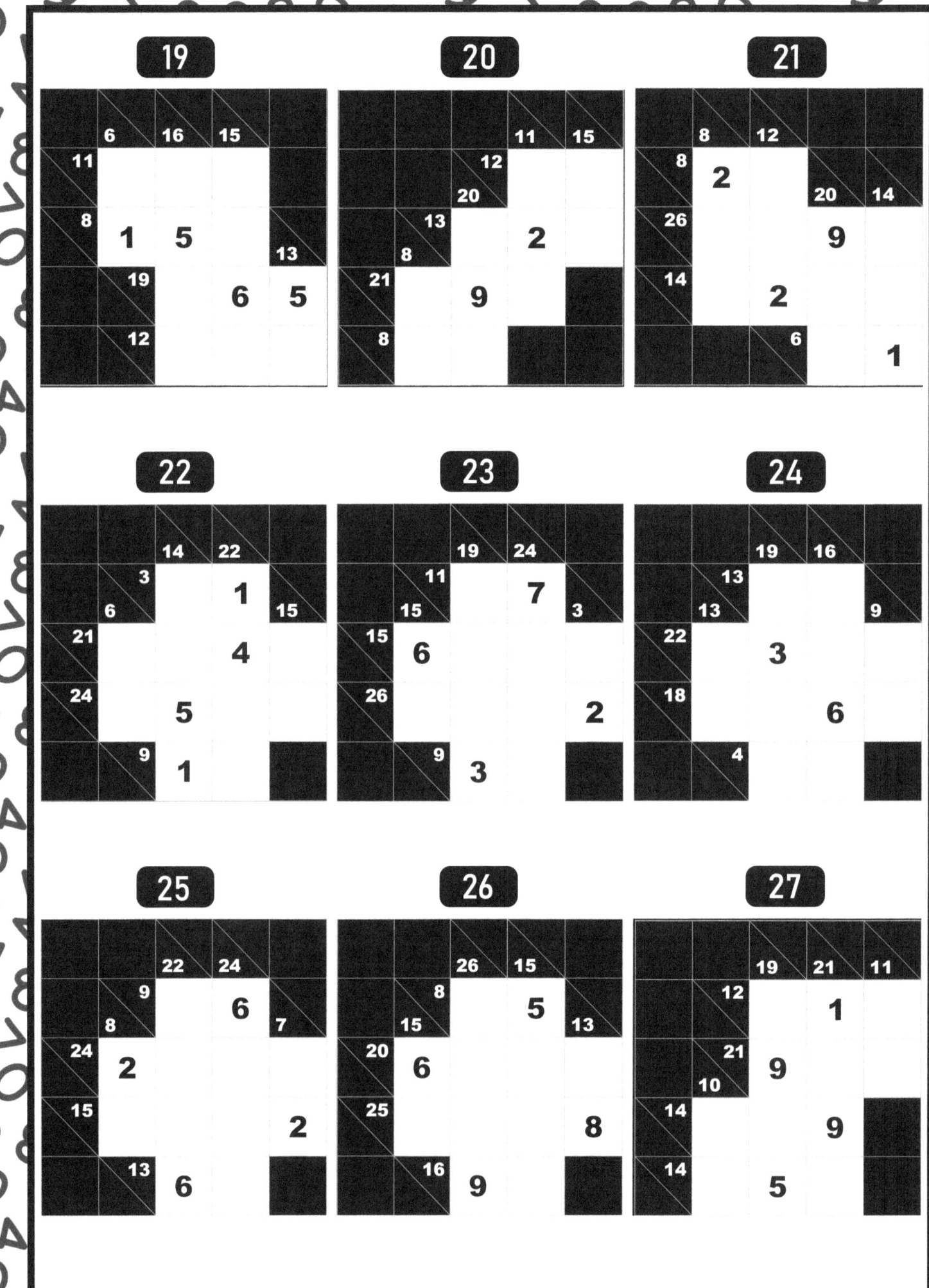

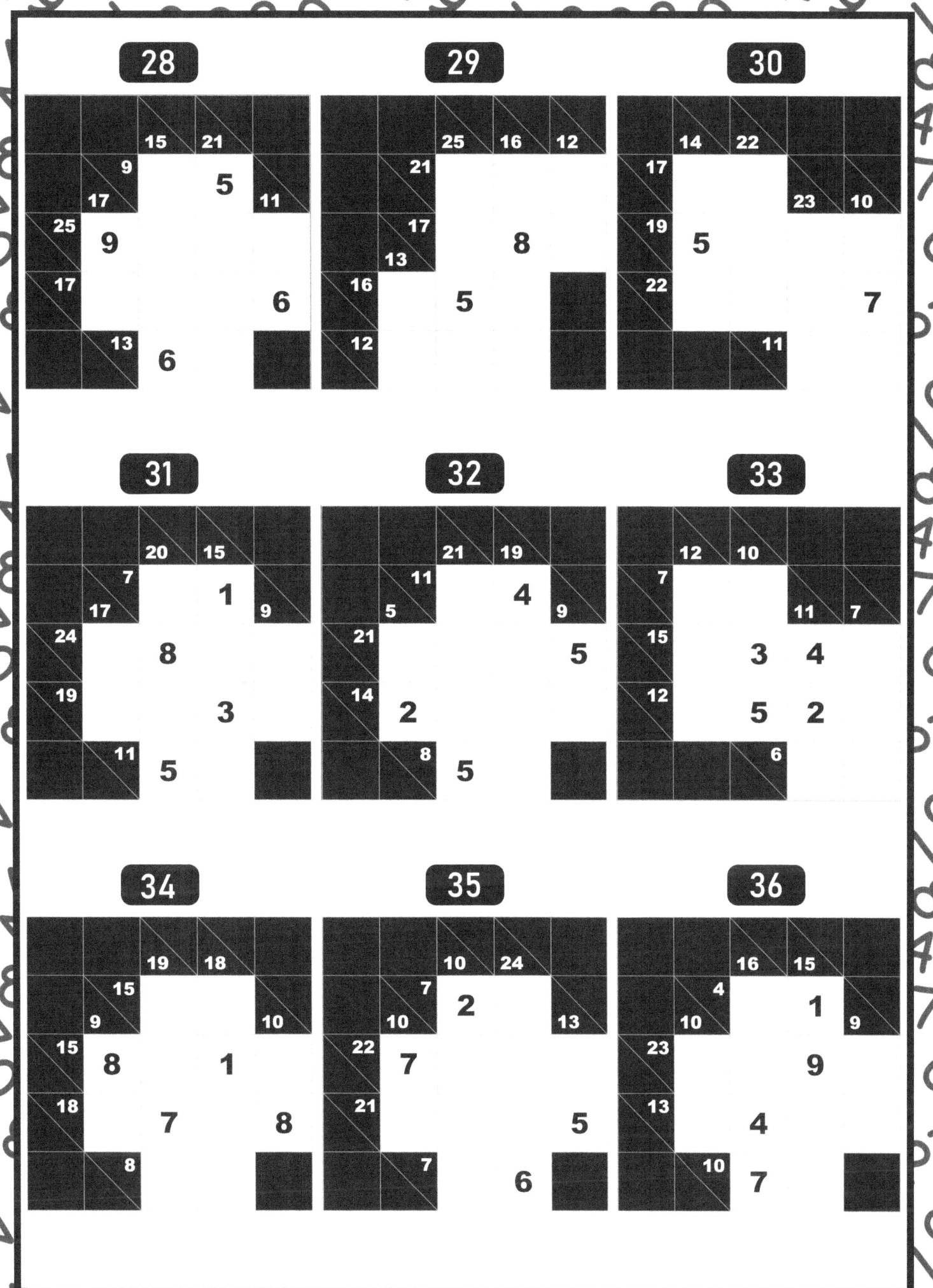

KAKURO

PUZZLE BOOK FOR ADULTS

-EASY-

6X6

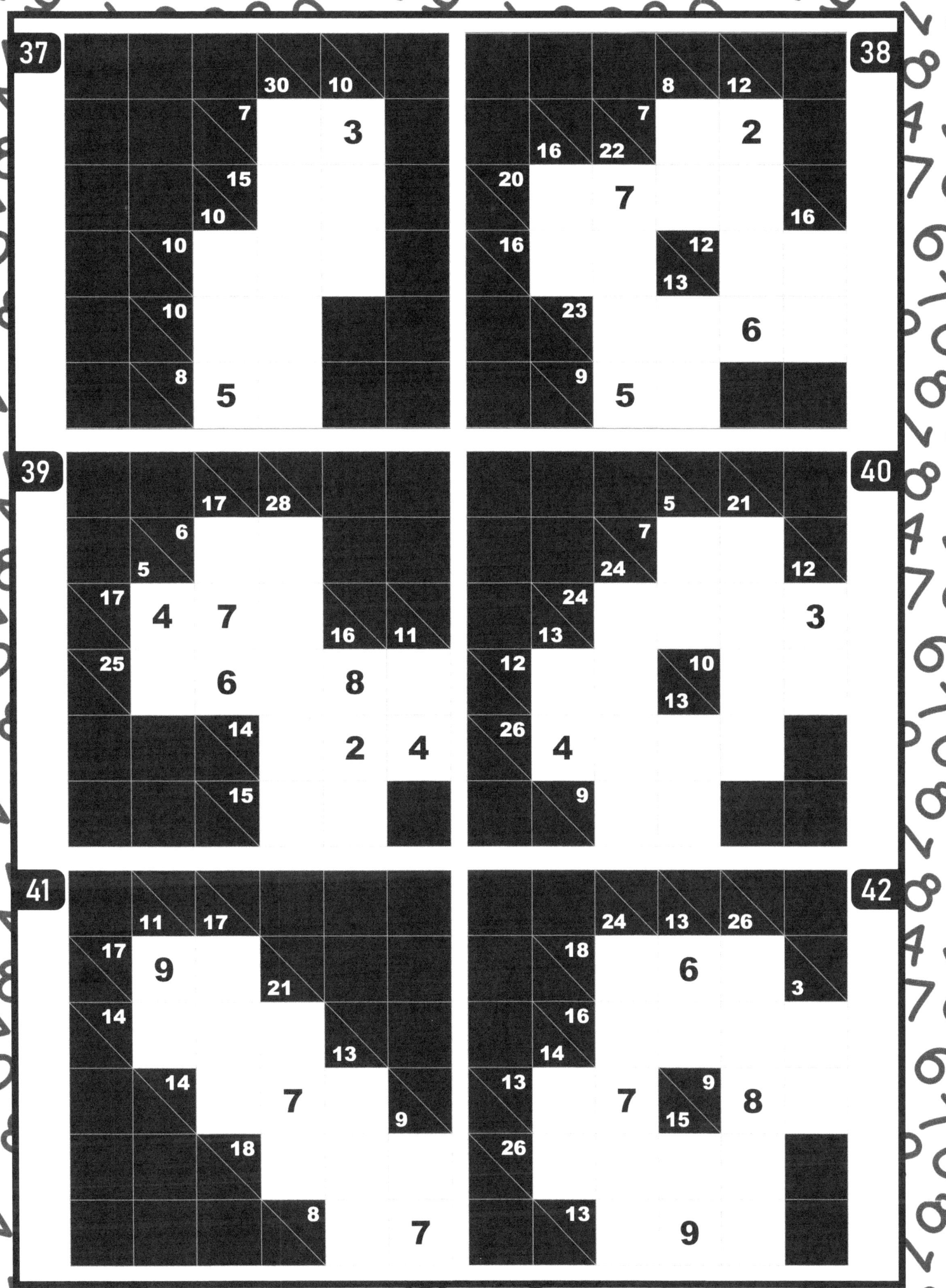

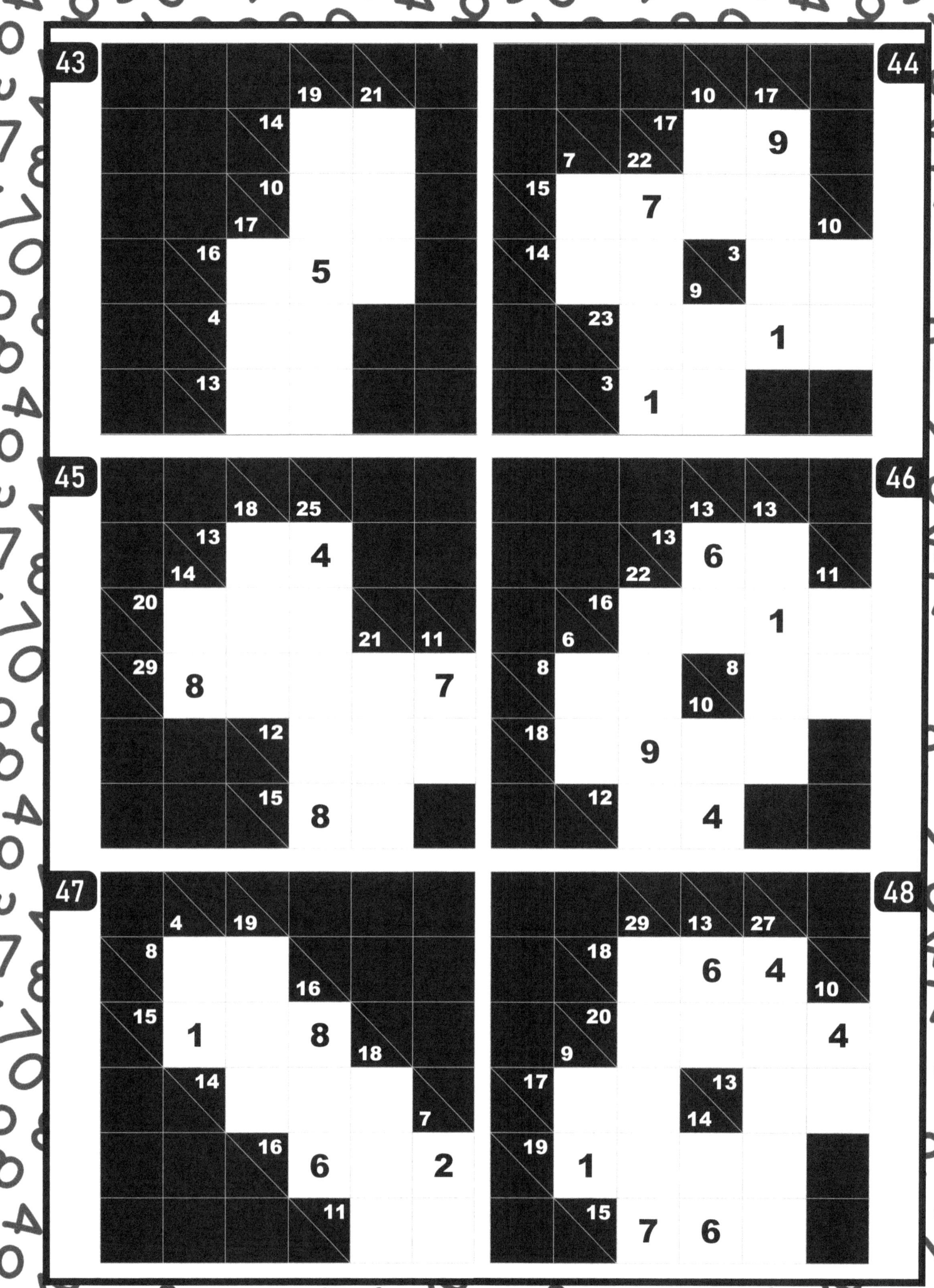

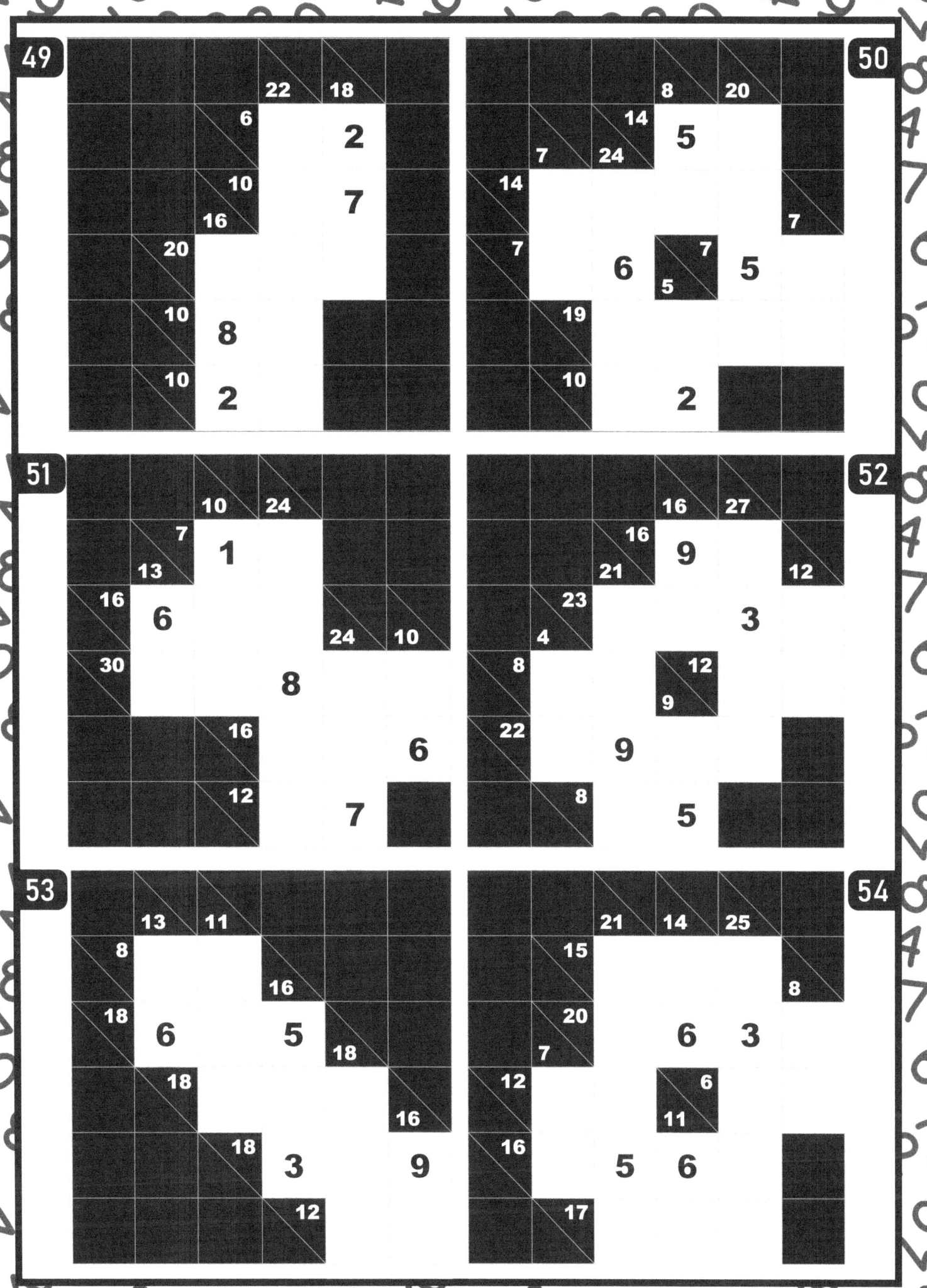

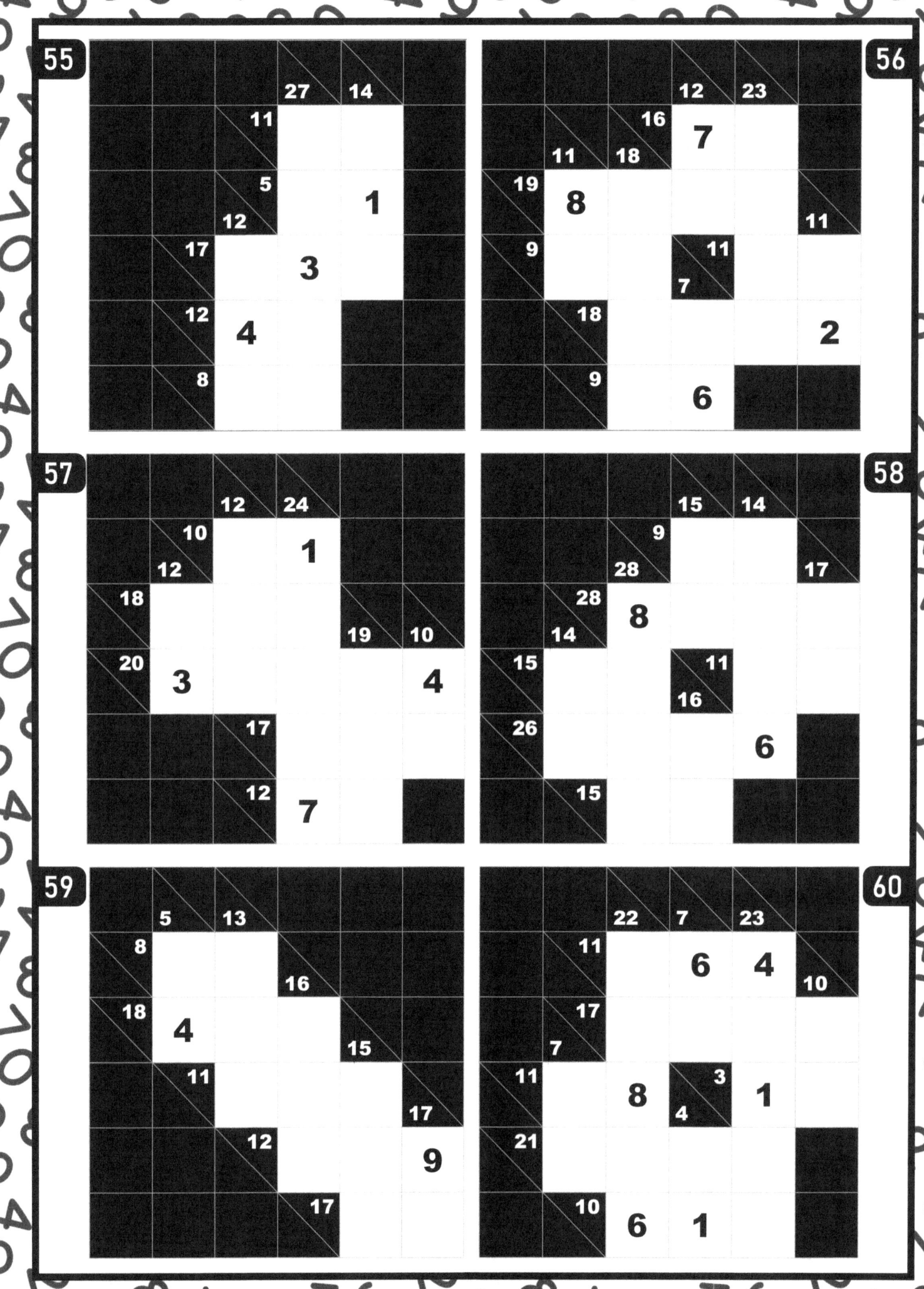

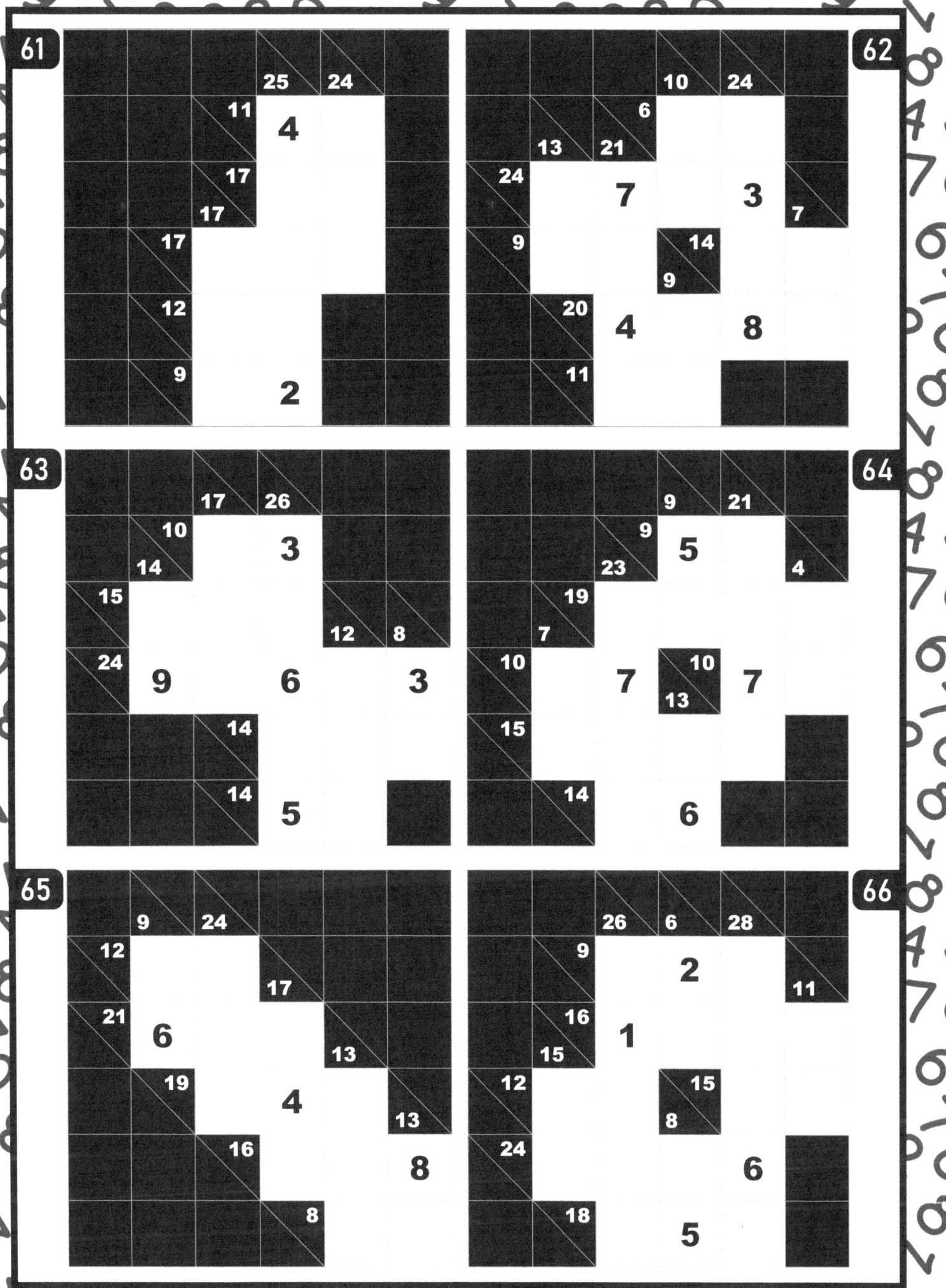

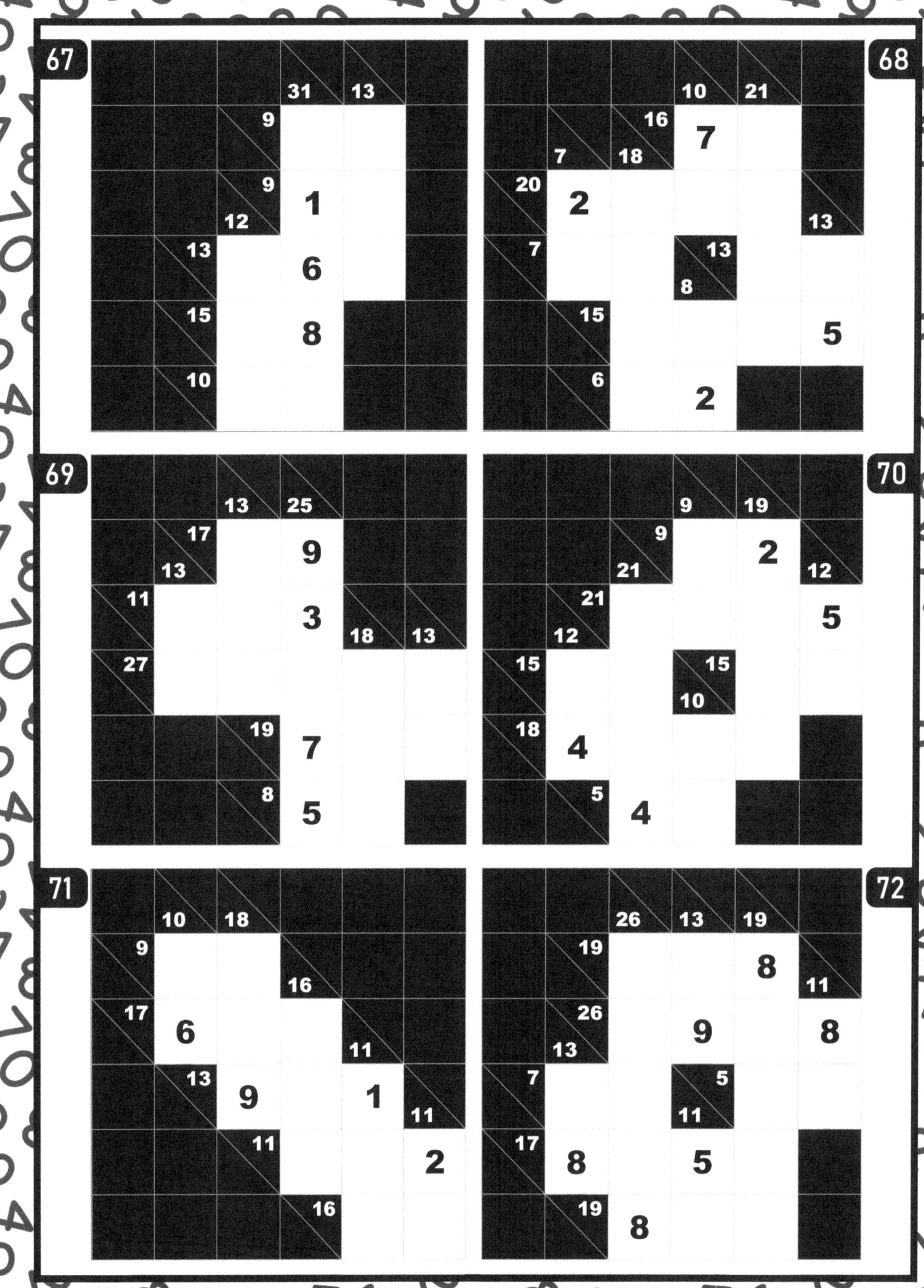

KAKURO

PUZZLE BOOK FOR ADULTS

-EASY-

7X8

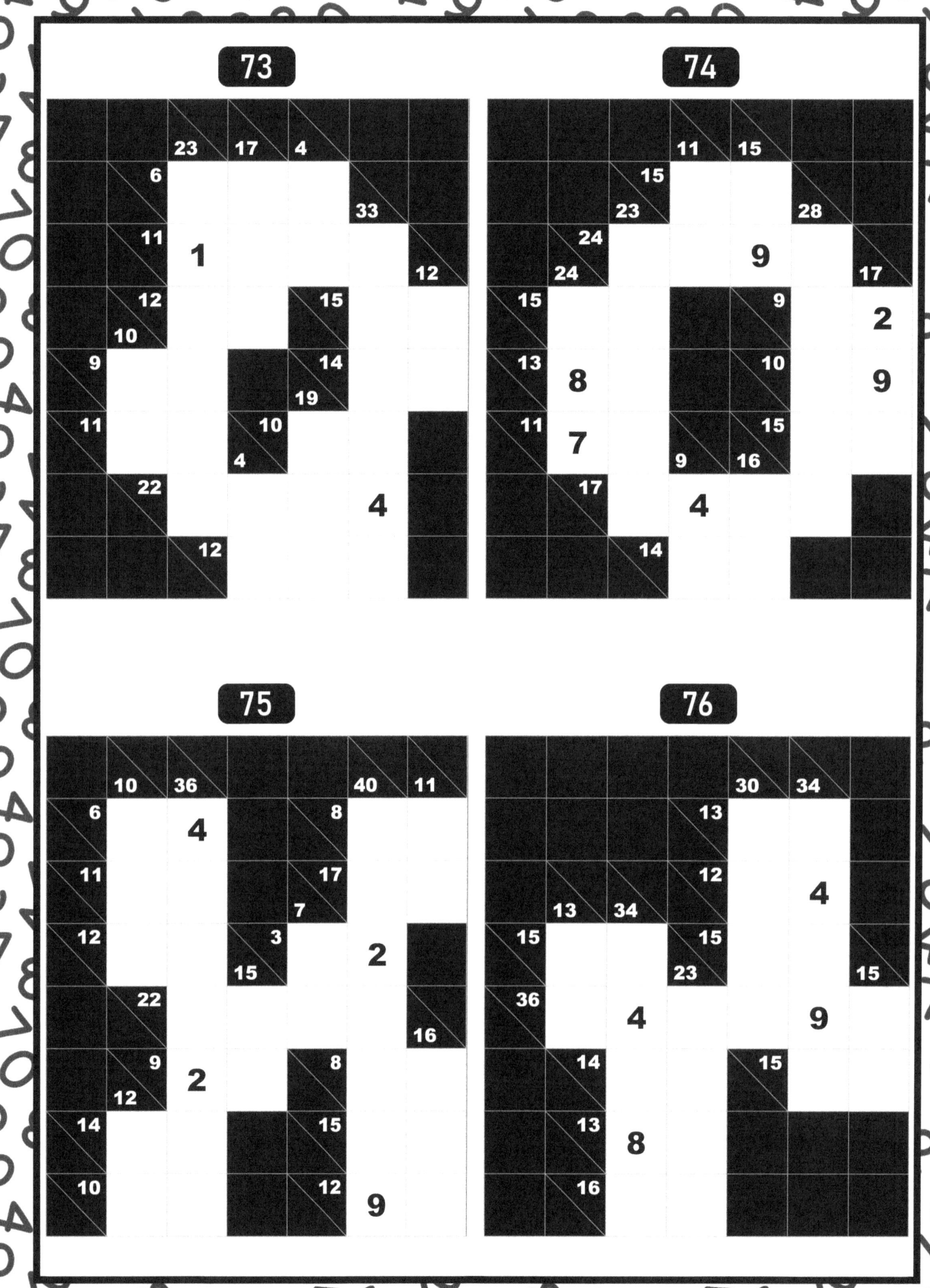

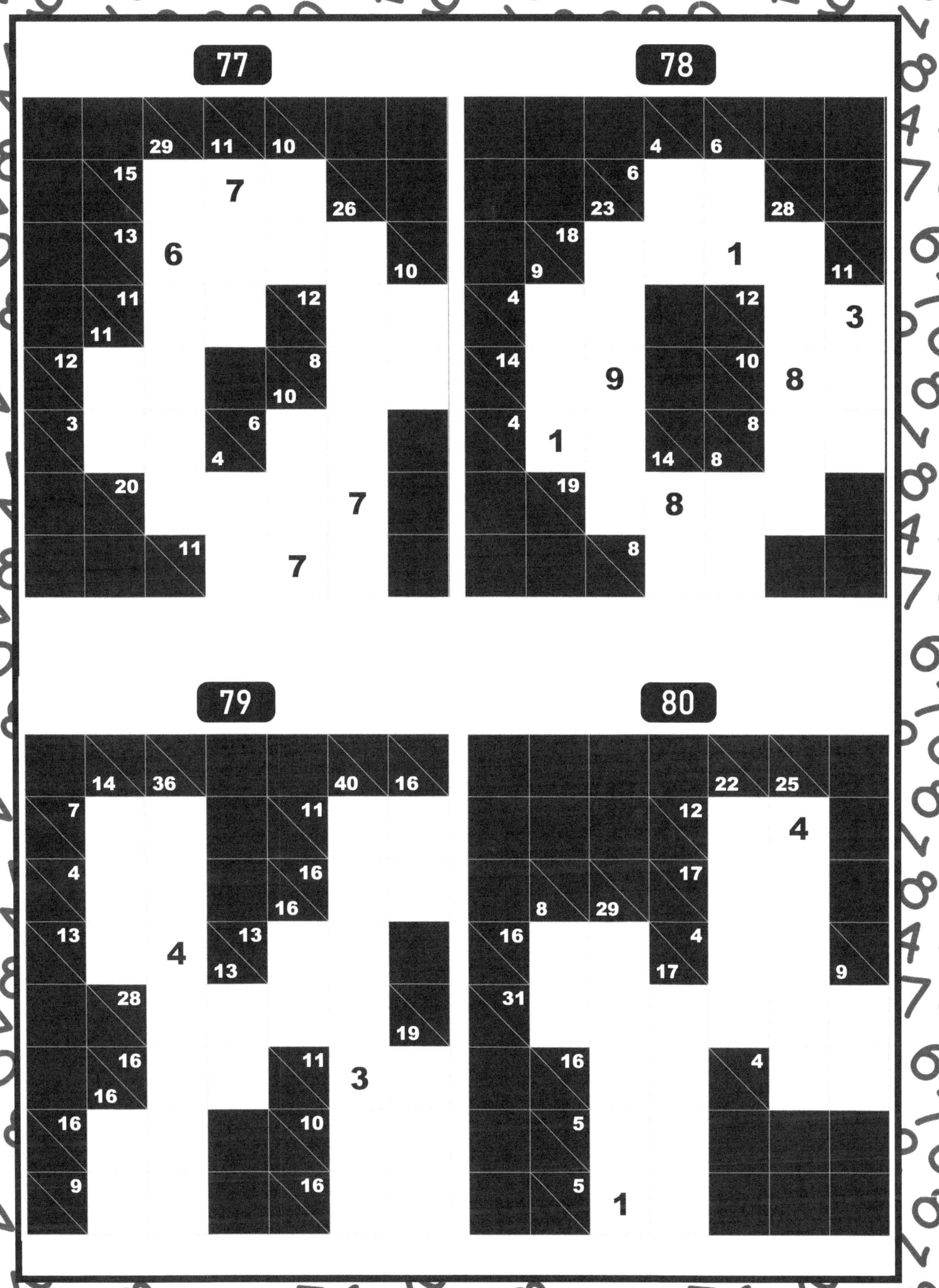

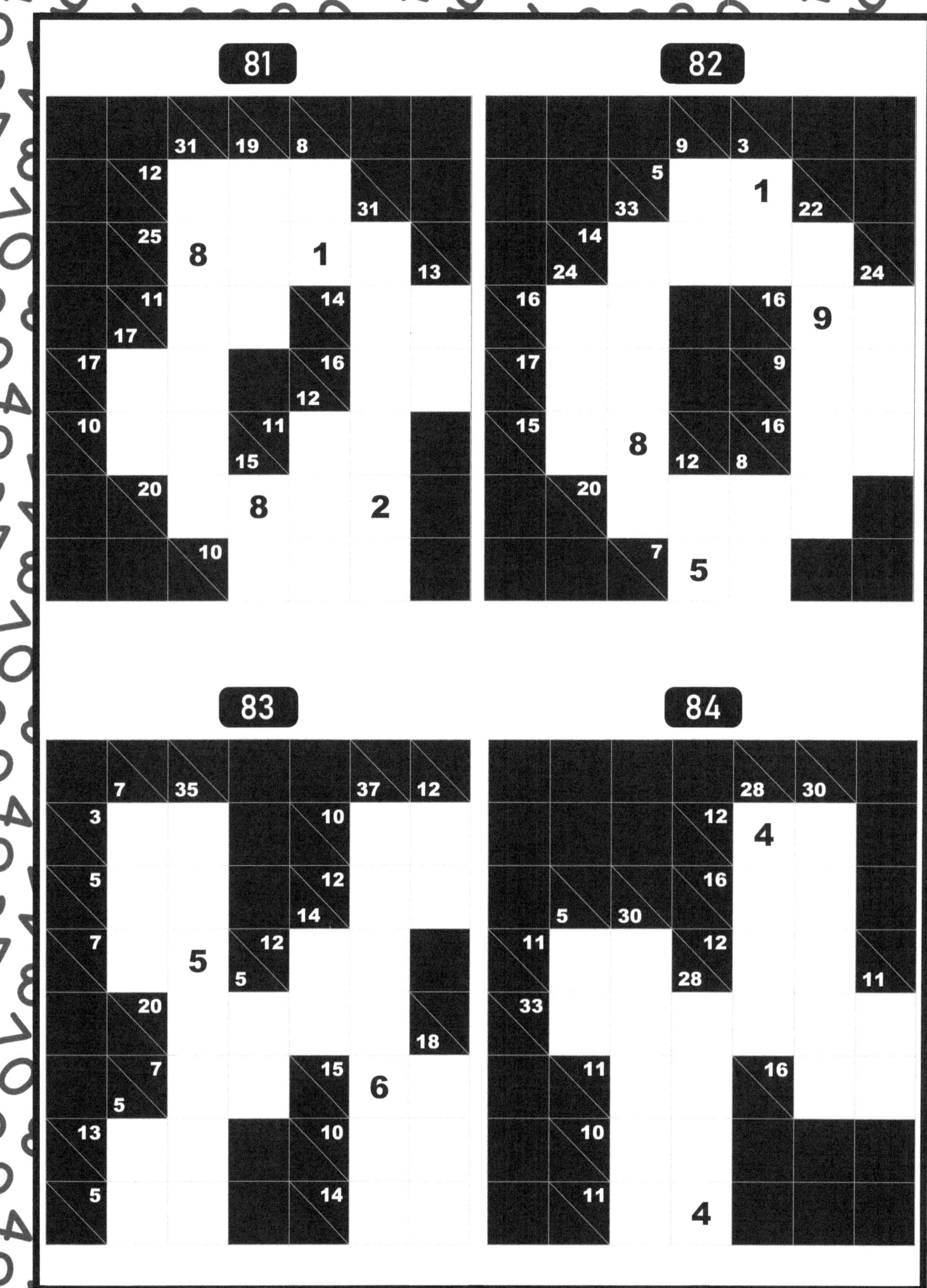

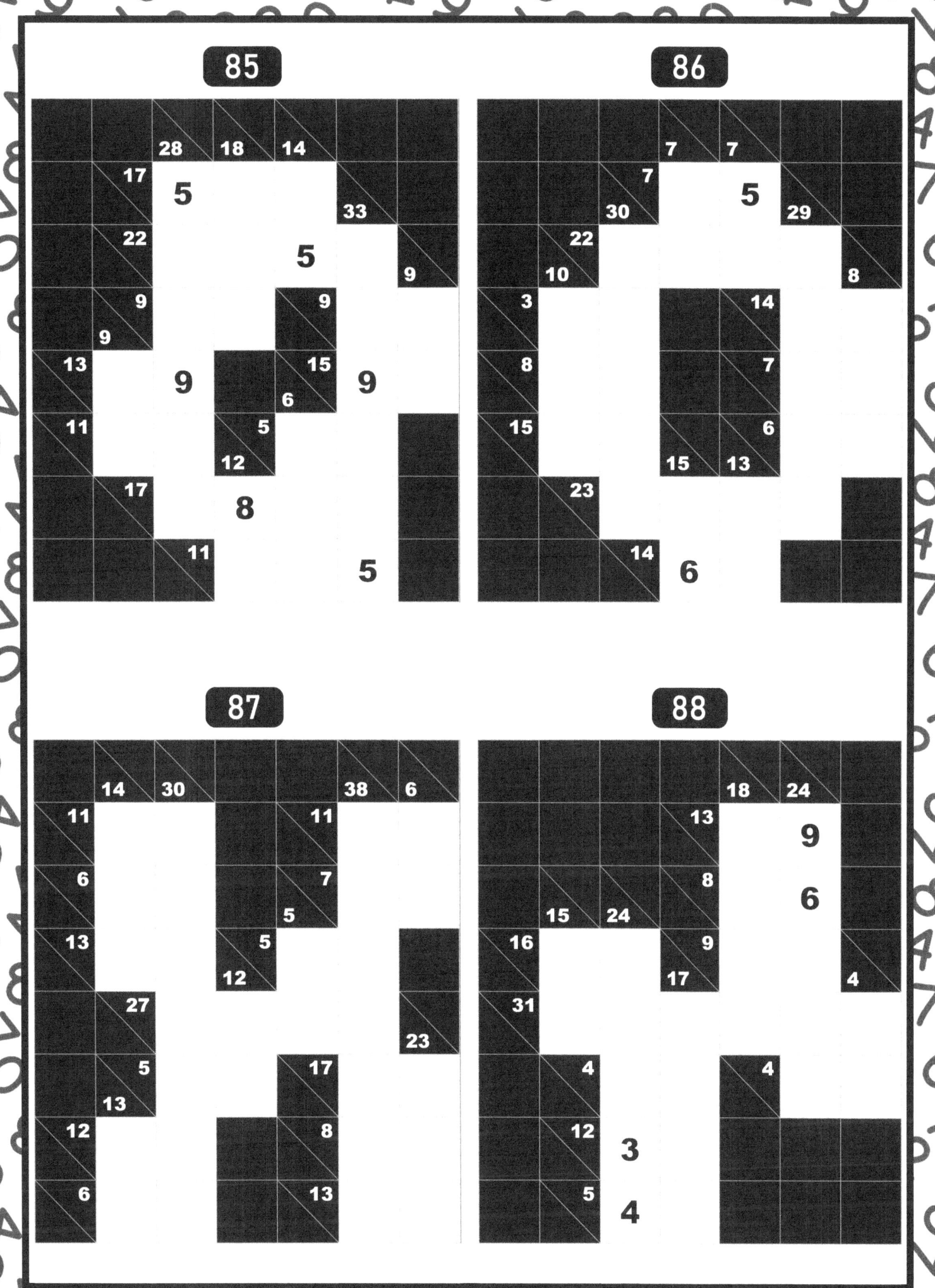

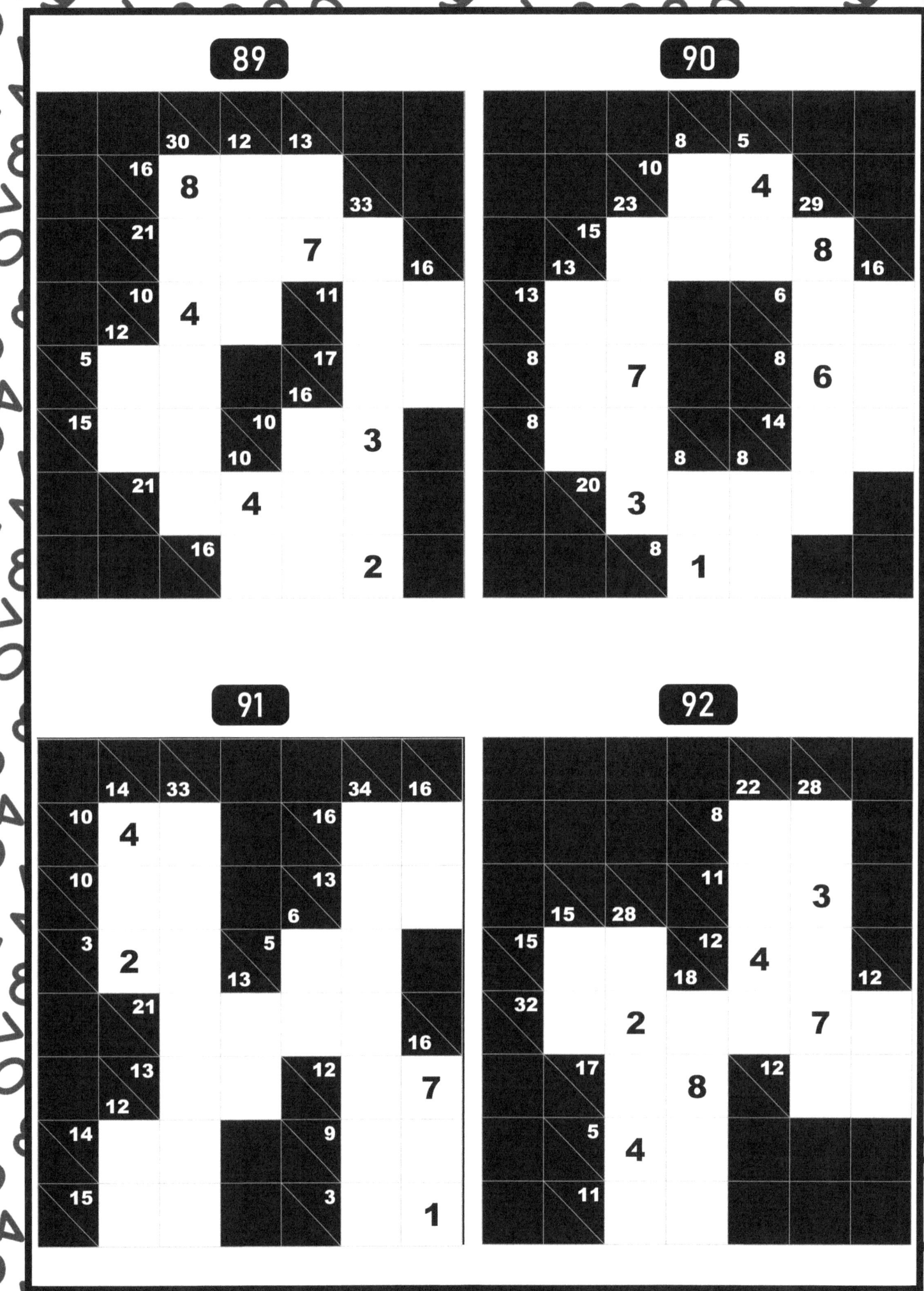

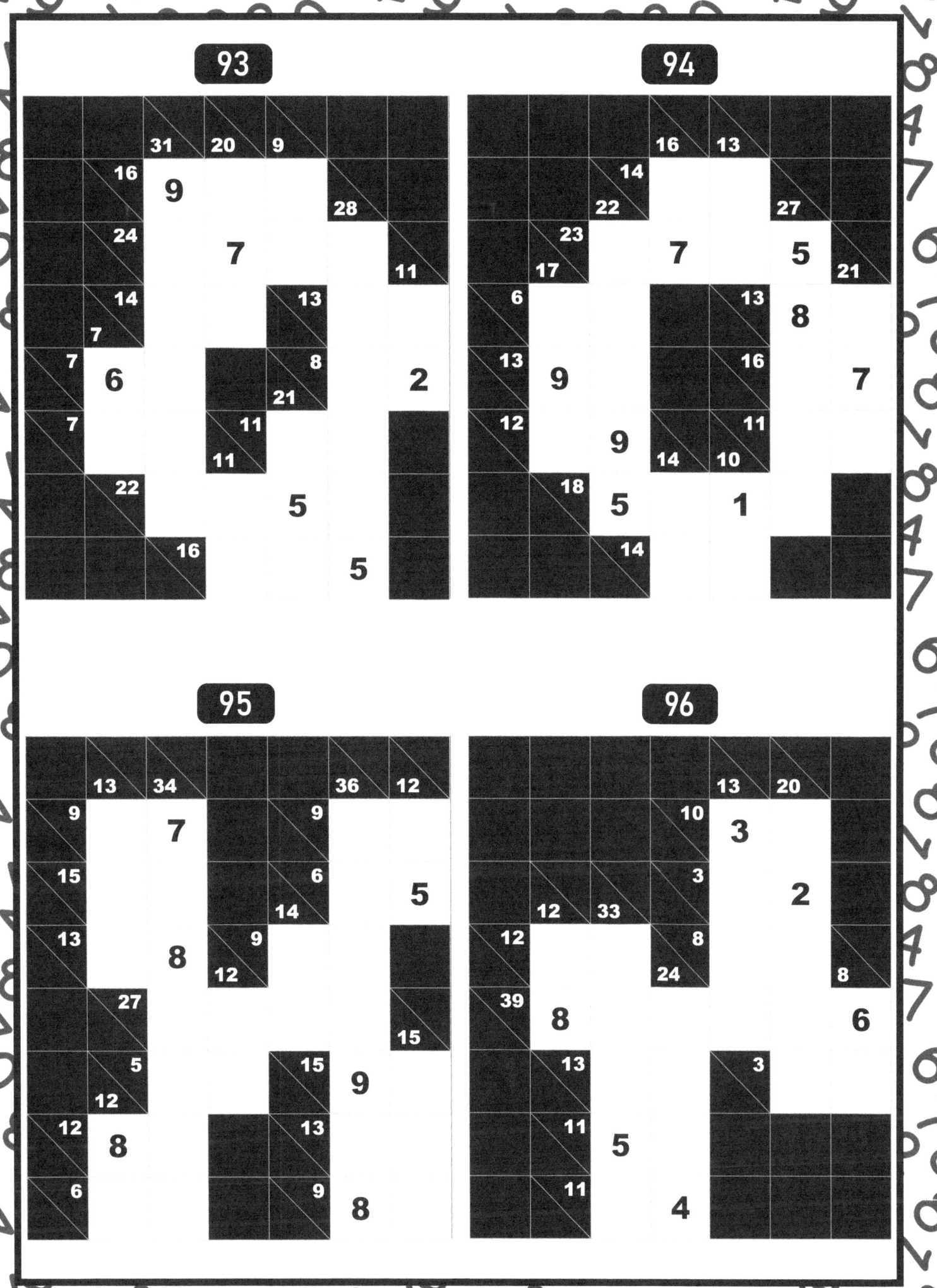

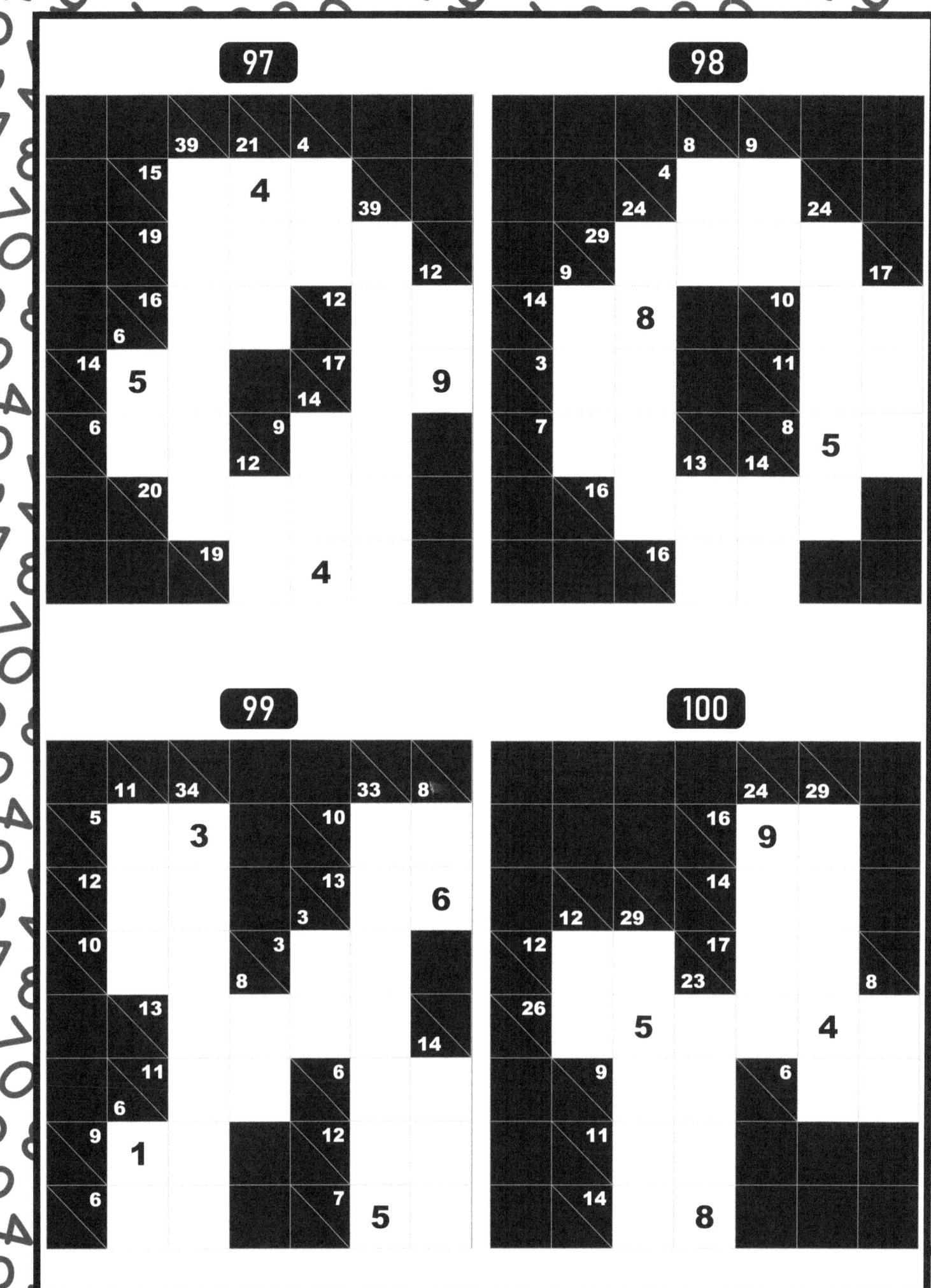

KAKURO

PUZZLE BOOK FOR ADULTS

-MEDIUM-

8X9

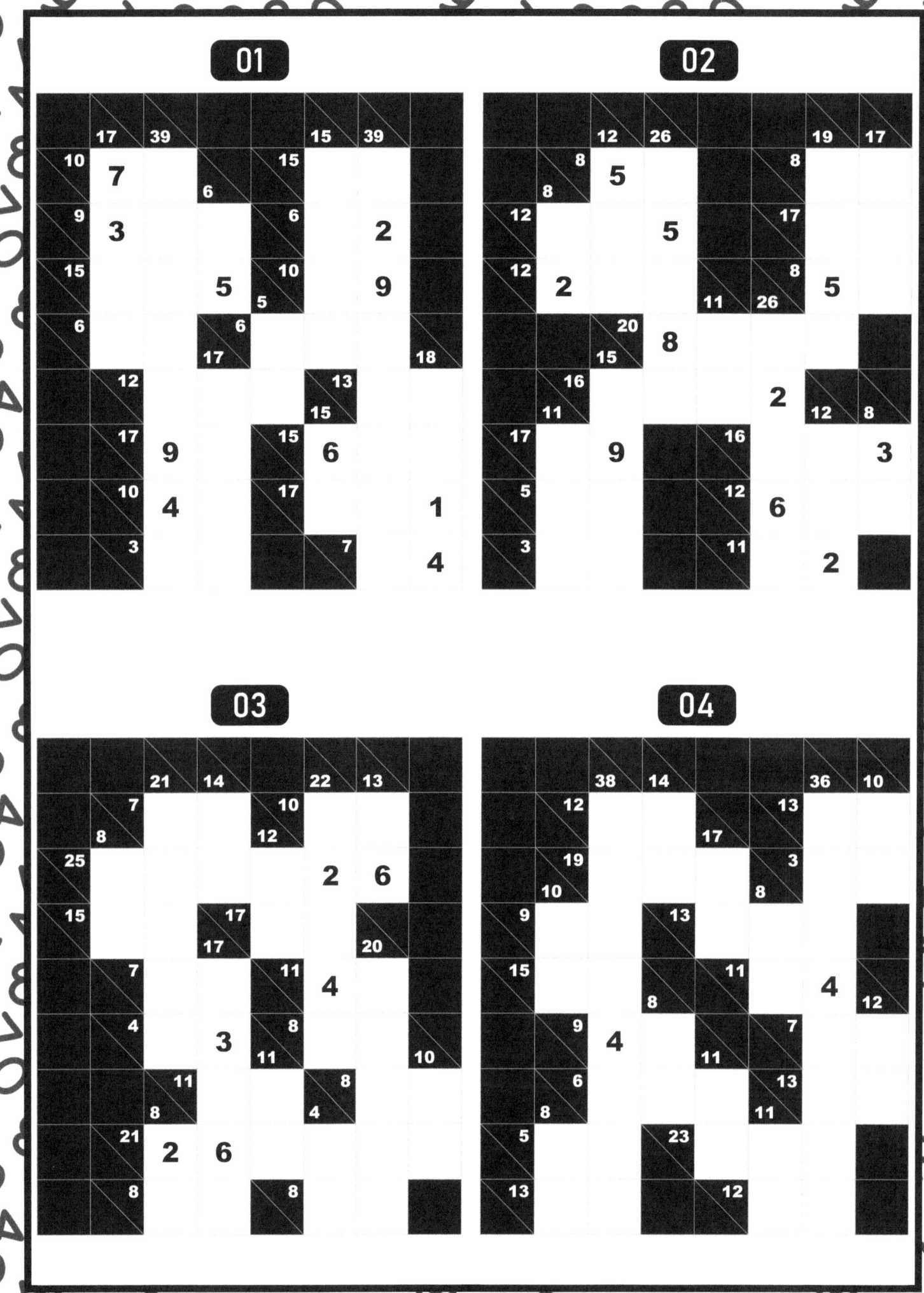

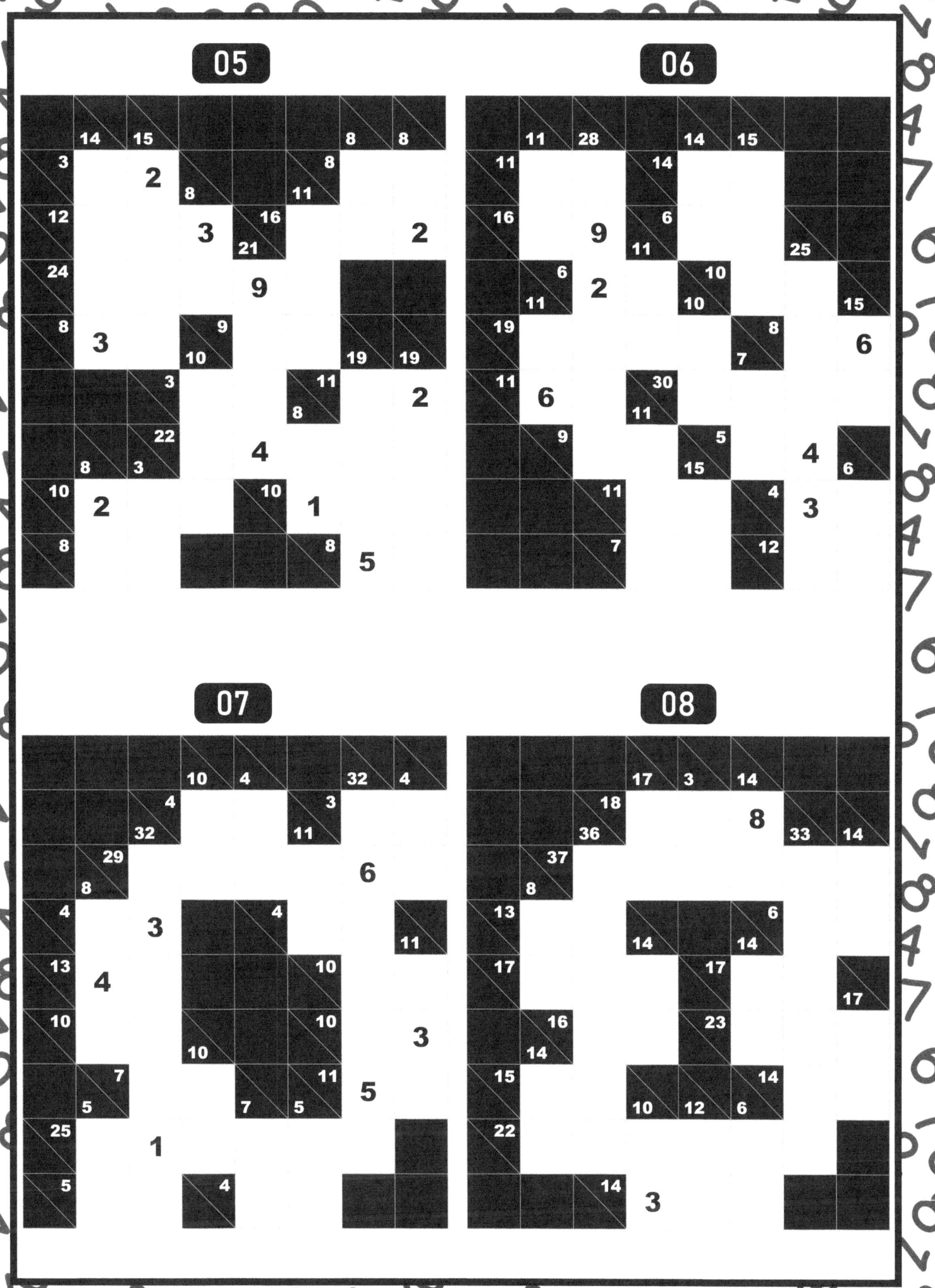

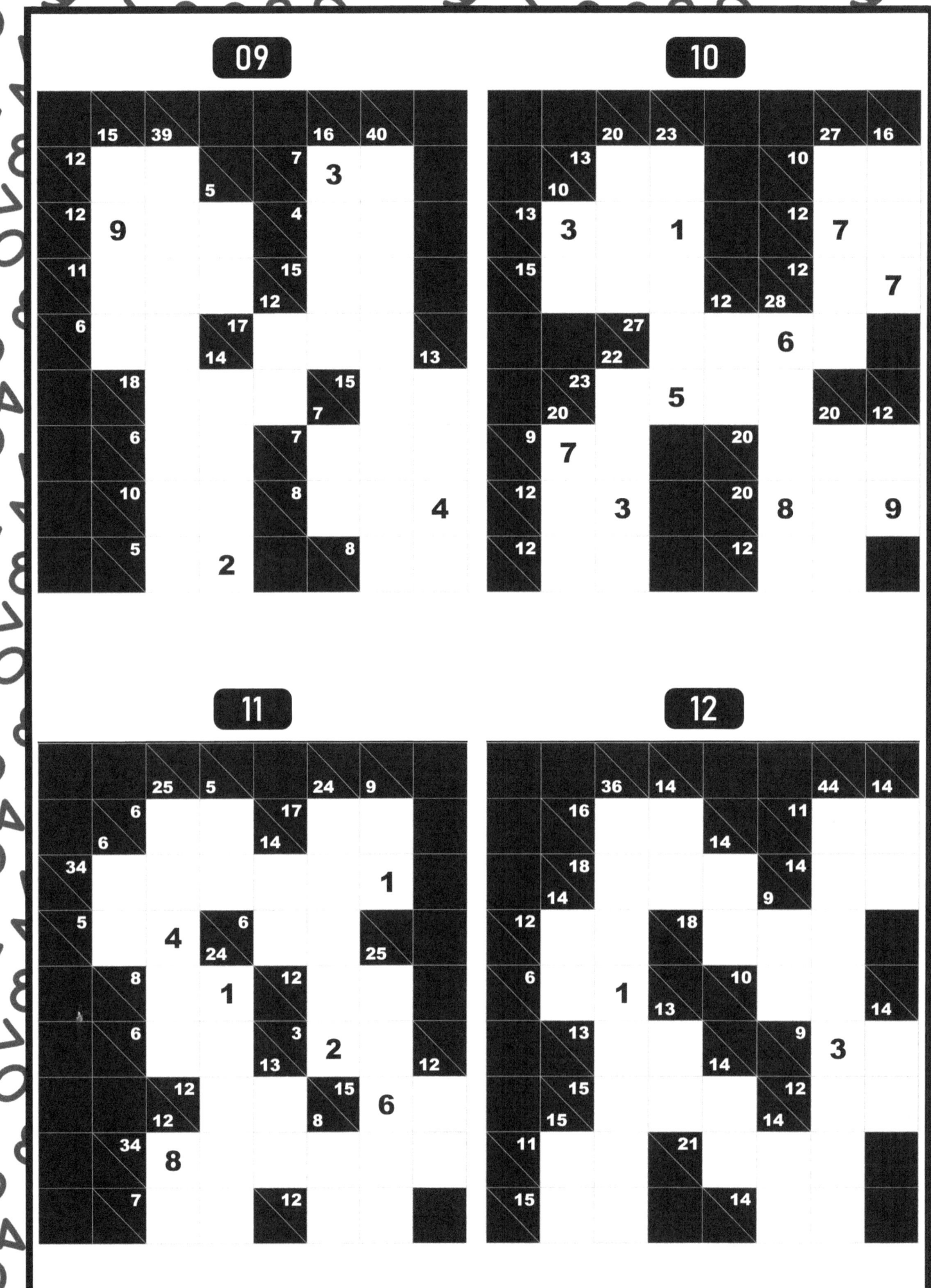

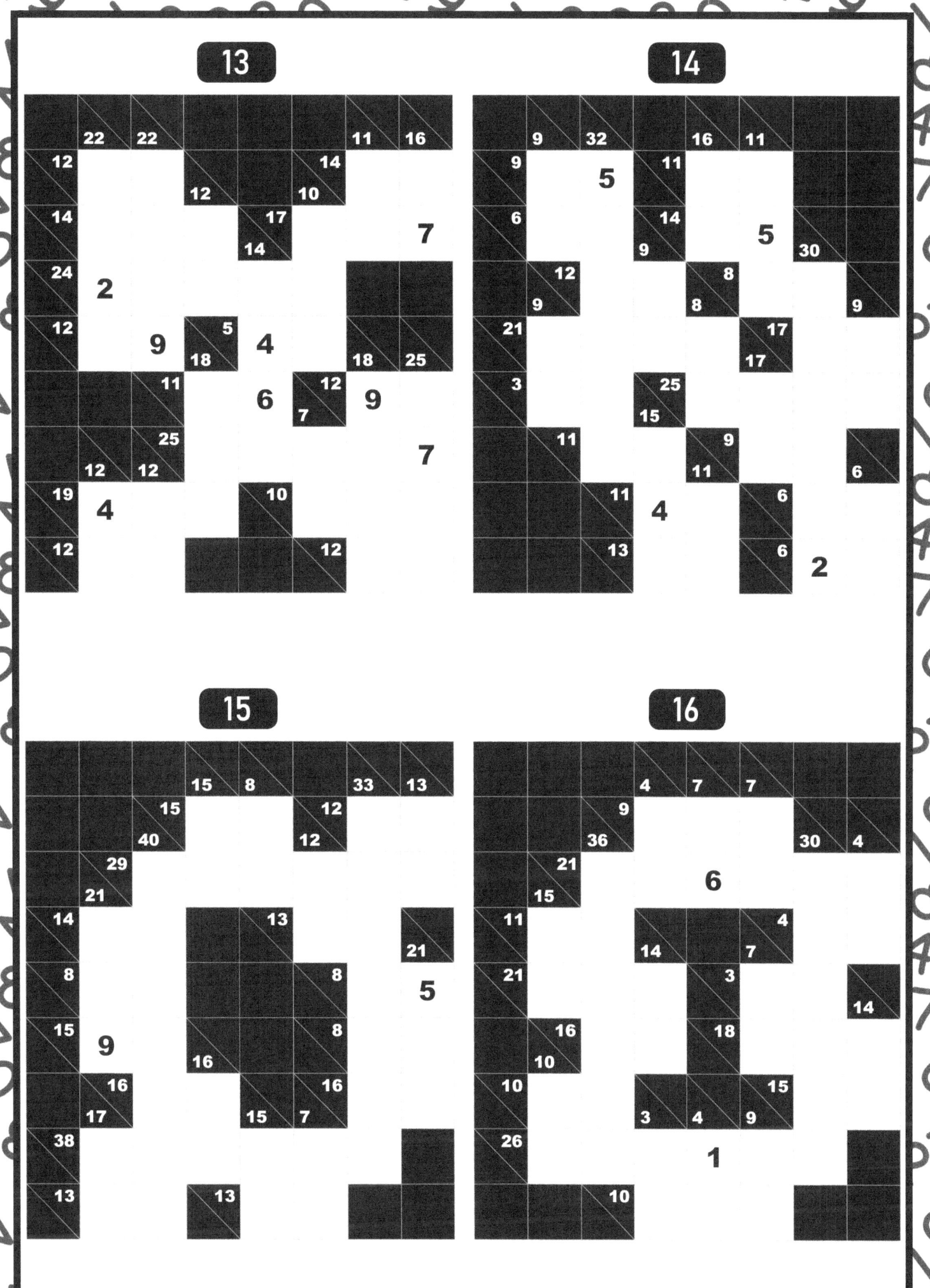

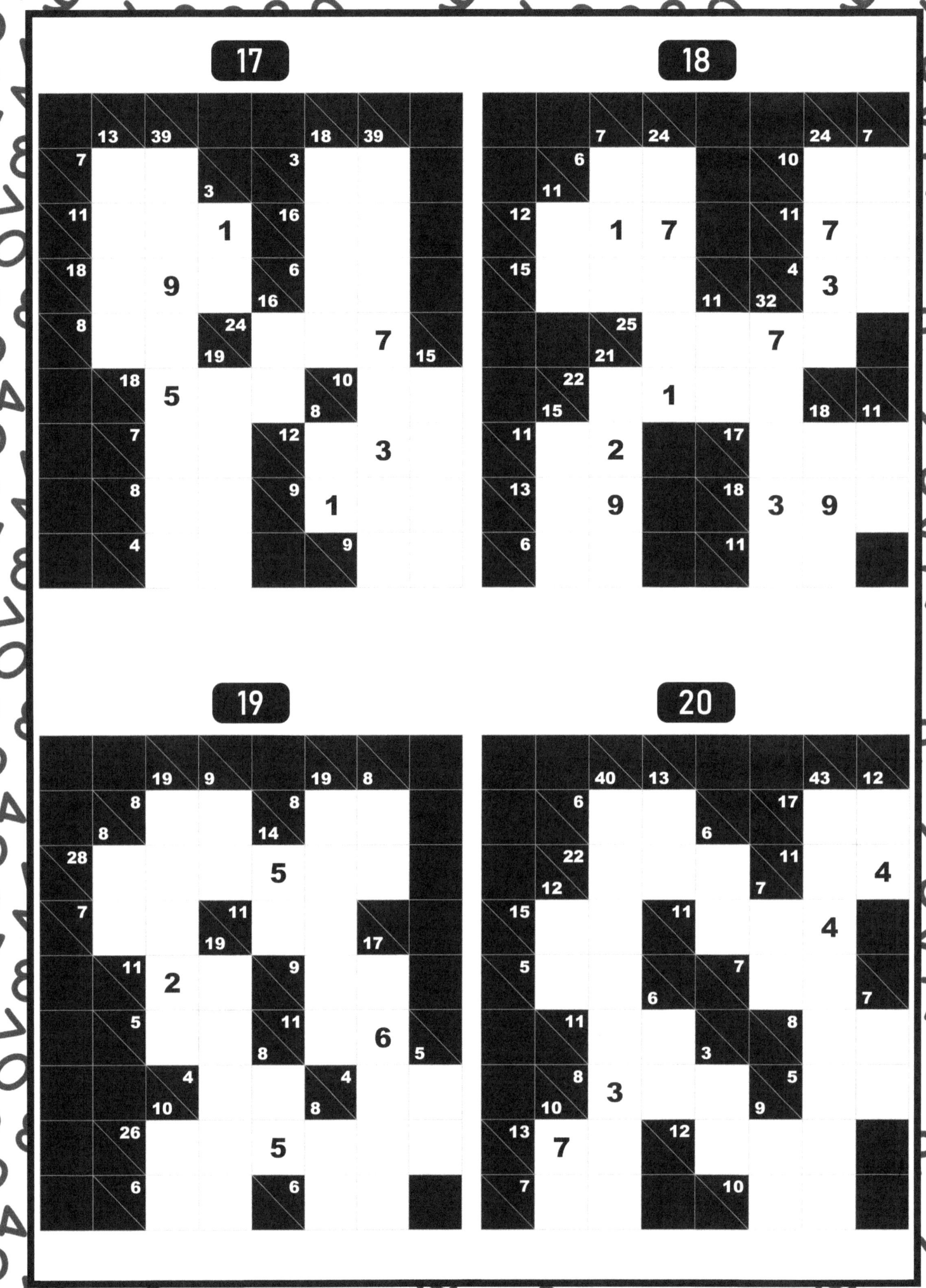

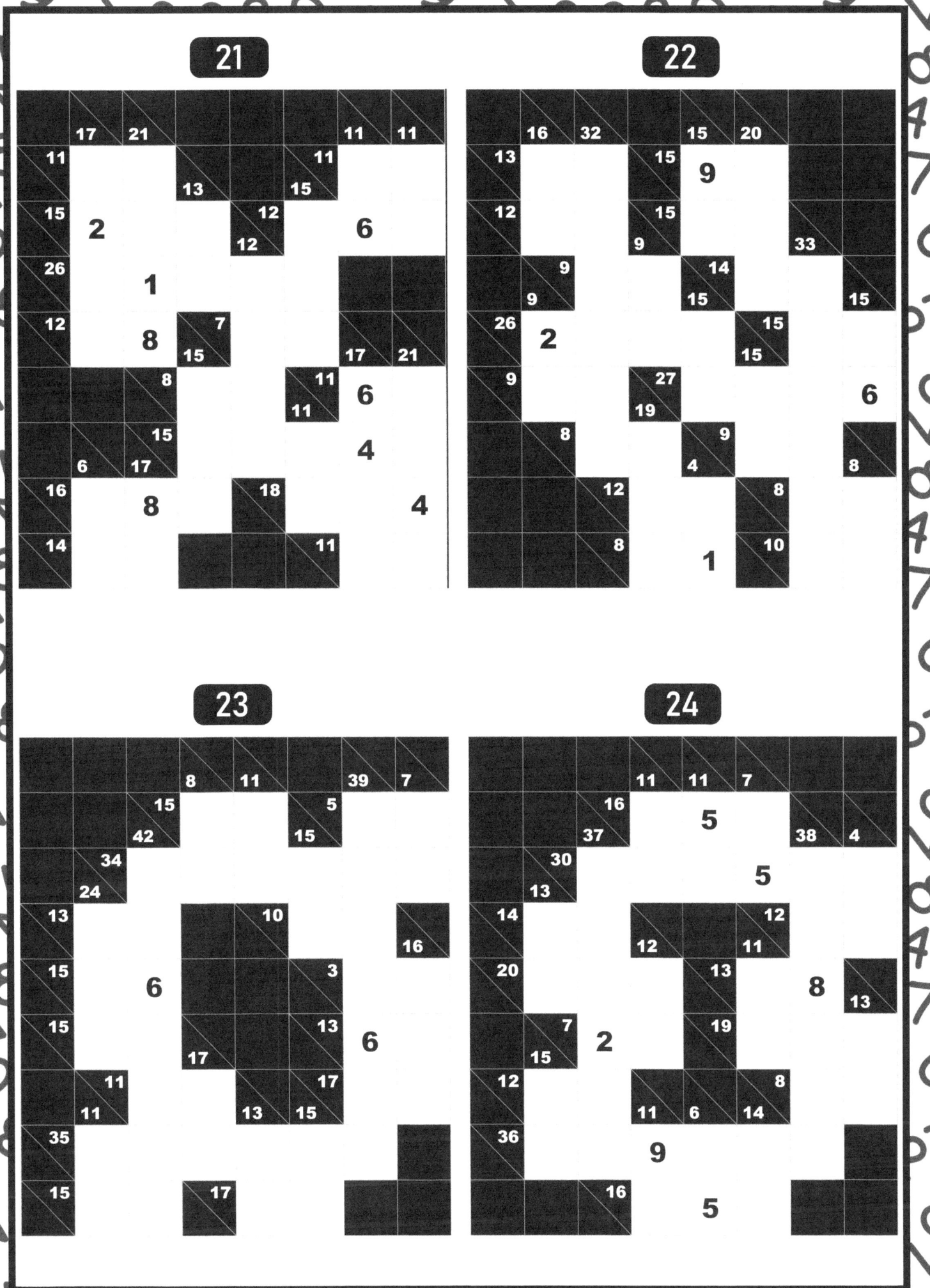

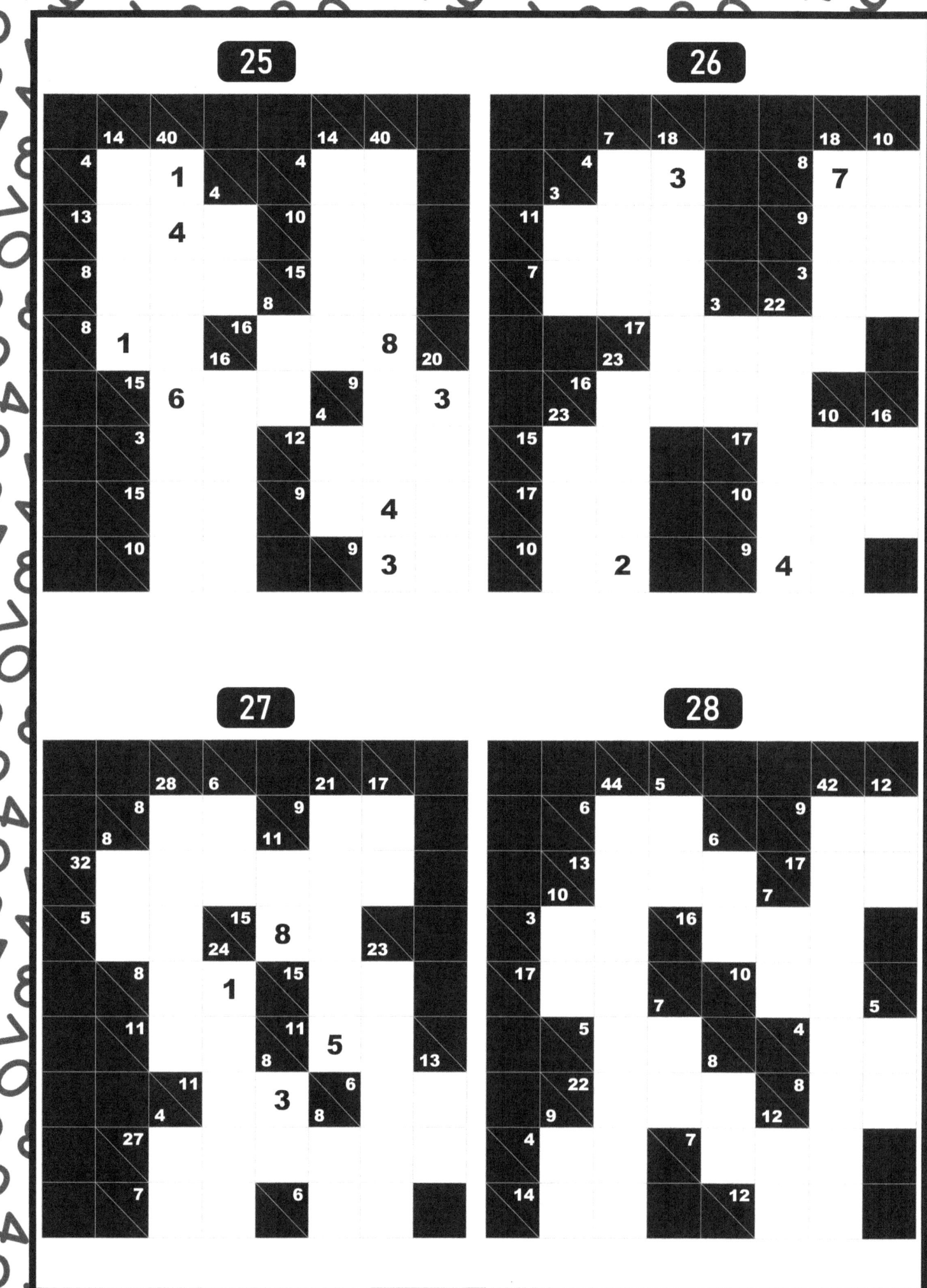

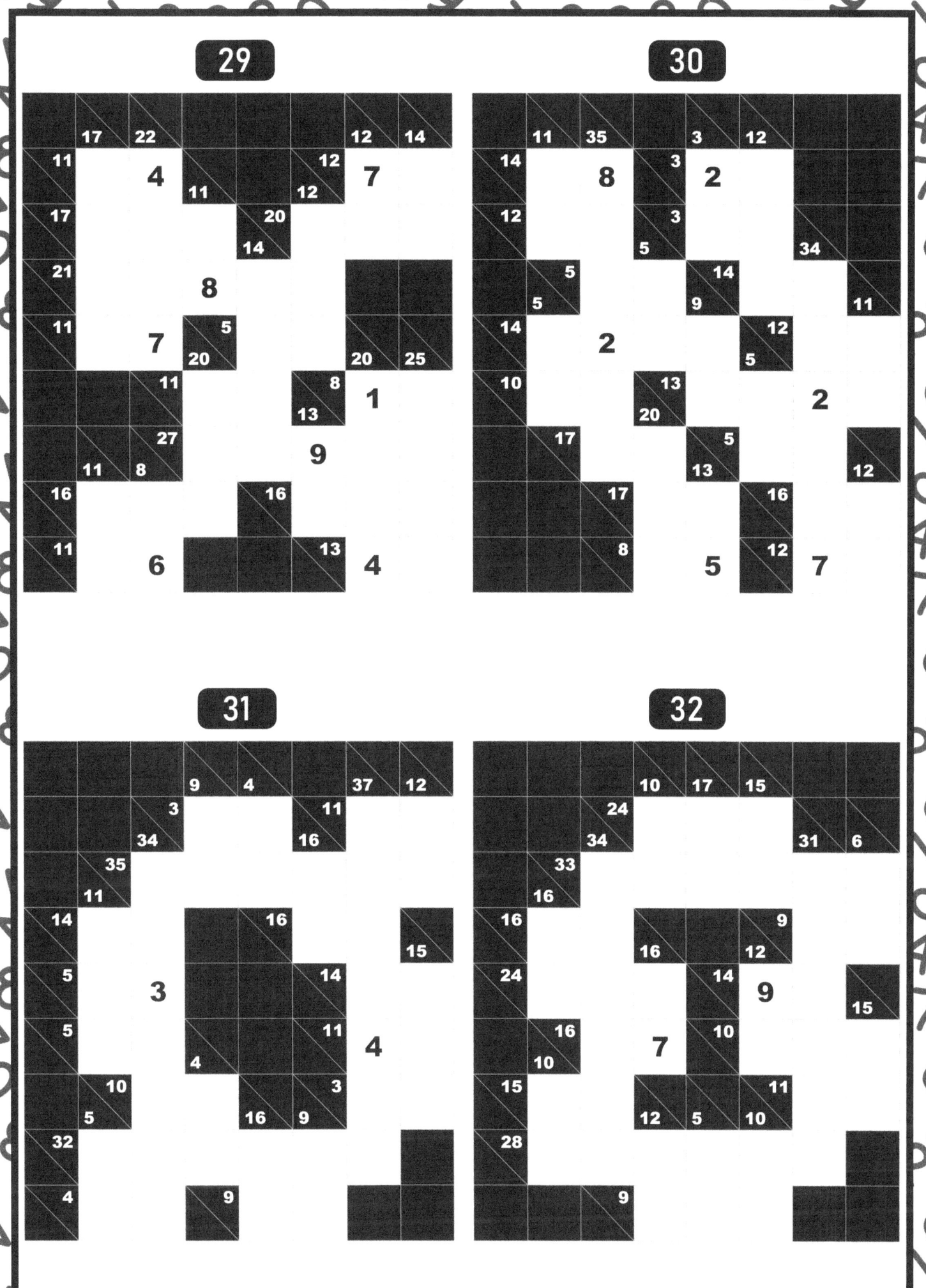

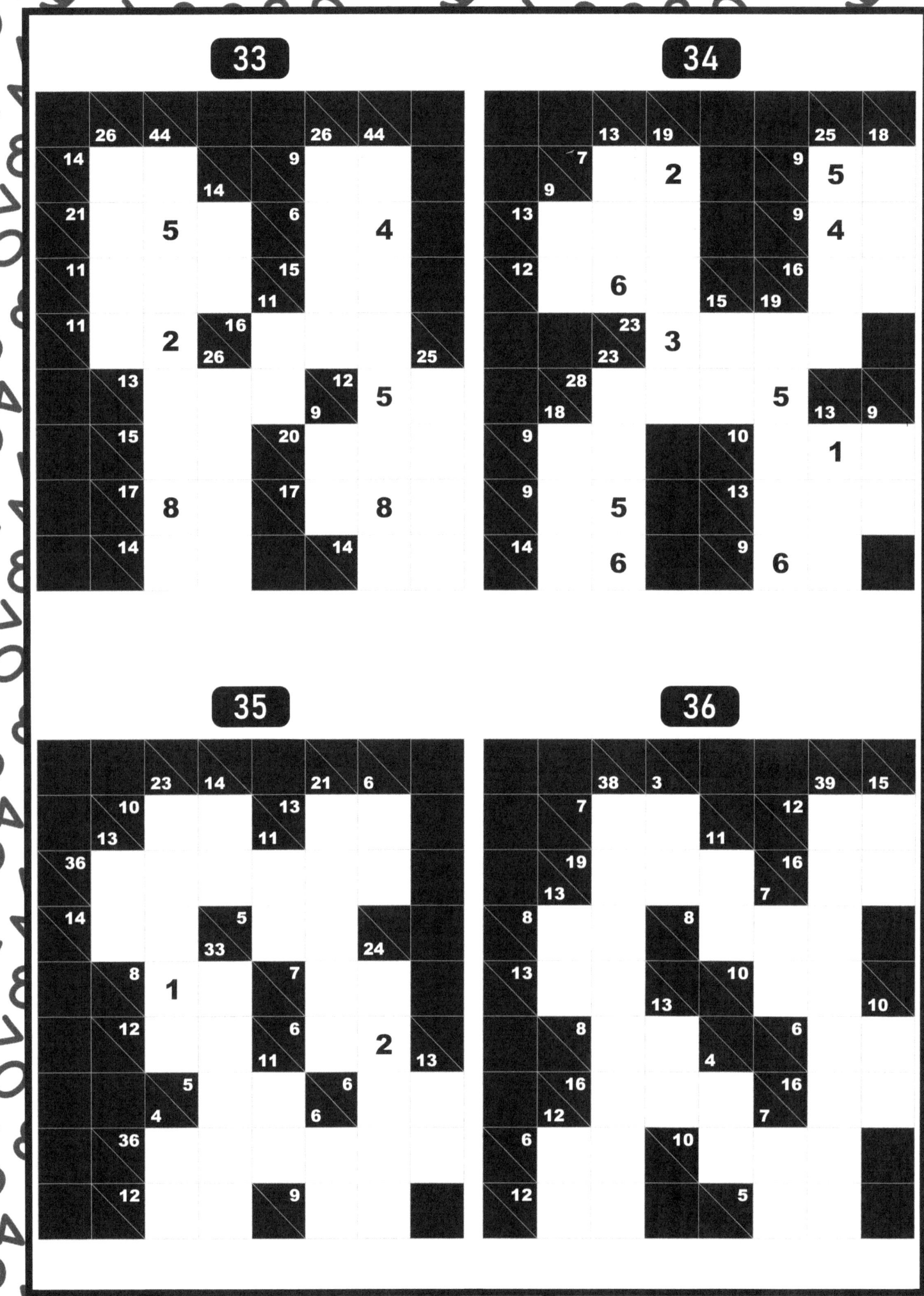

KAKURO

PUZZLE BOOK FOR ADULTS

-MEDIUM-

8X11

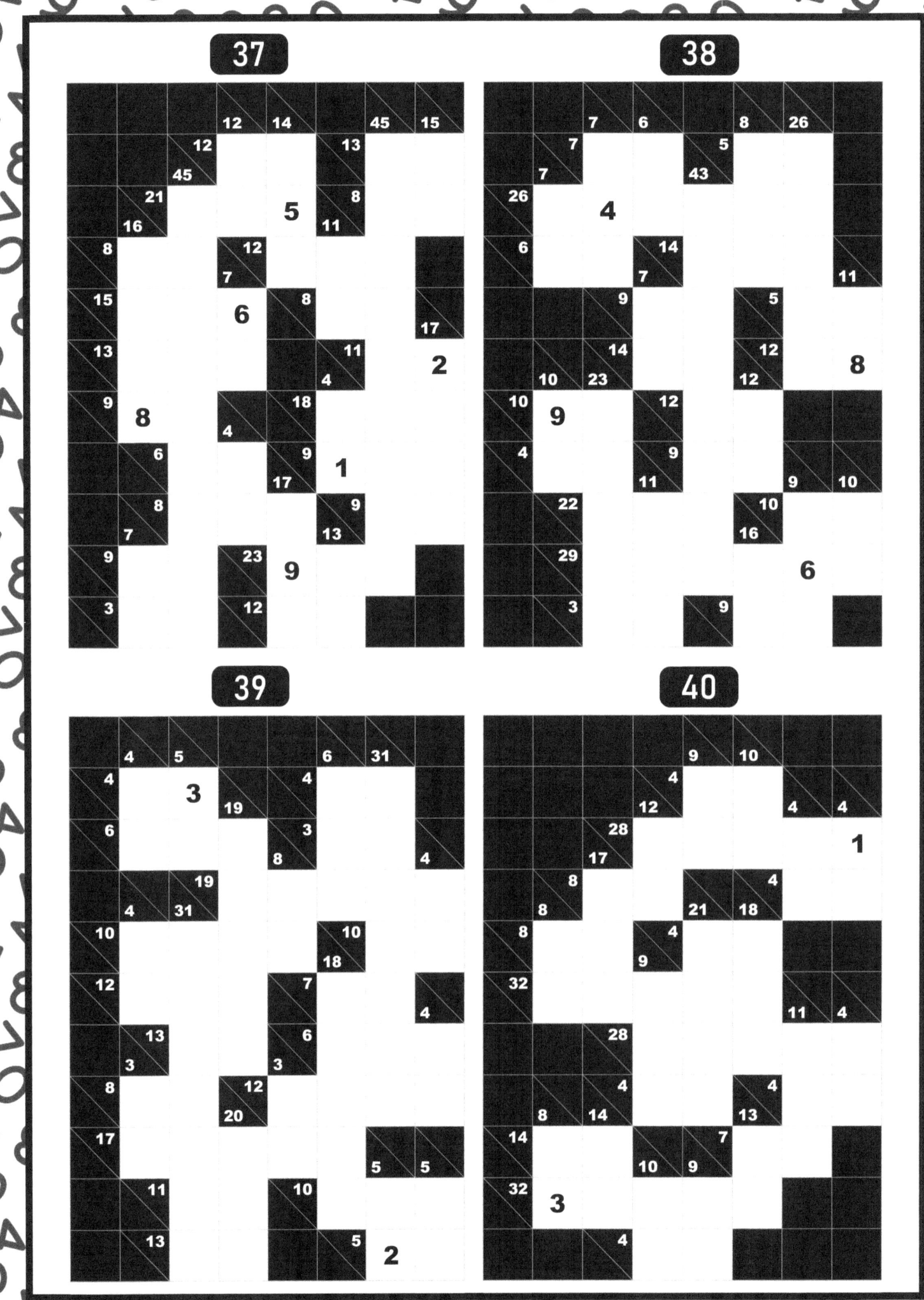

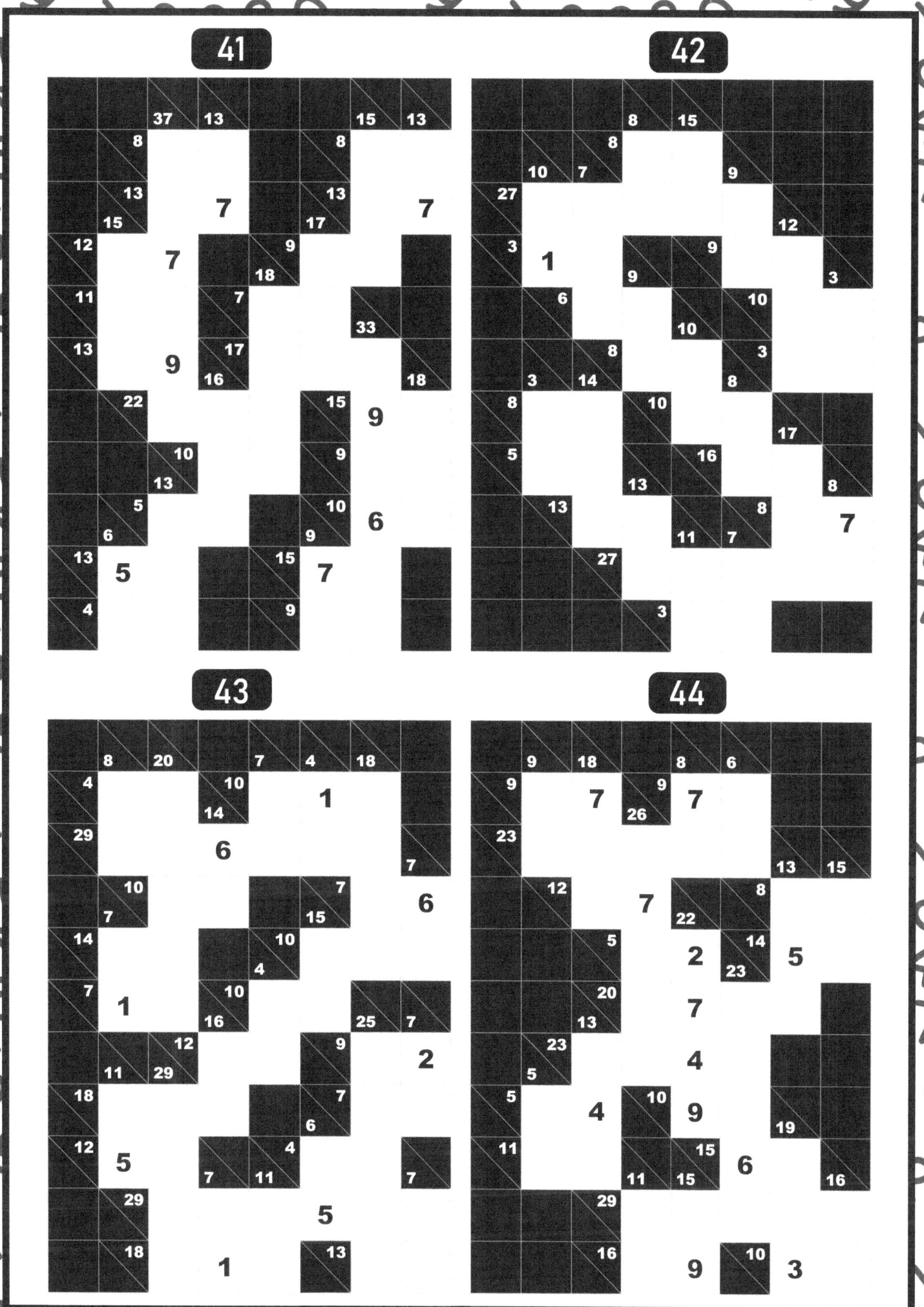

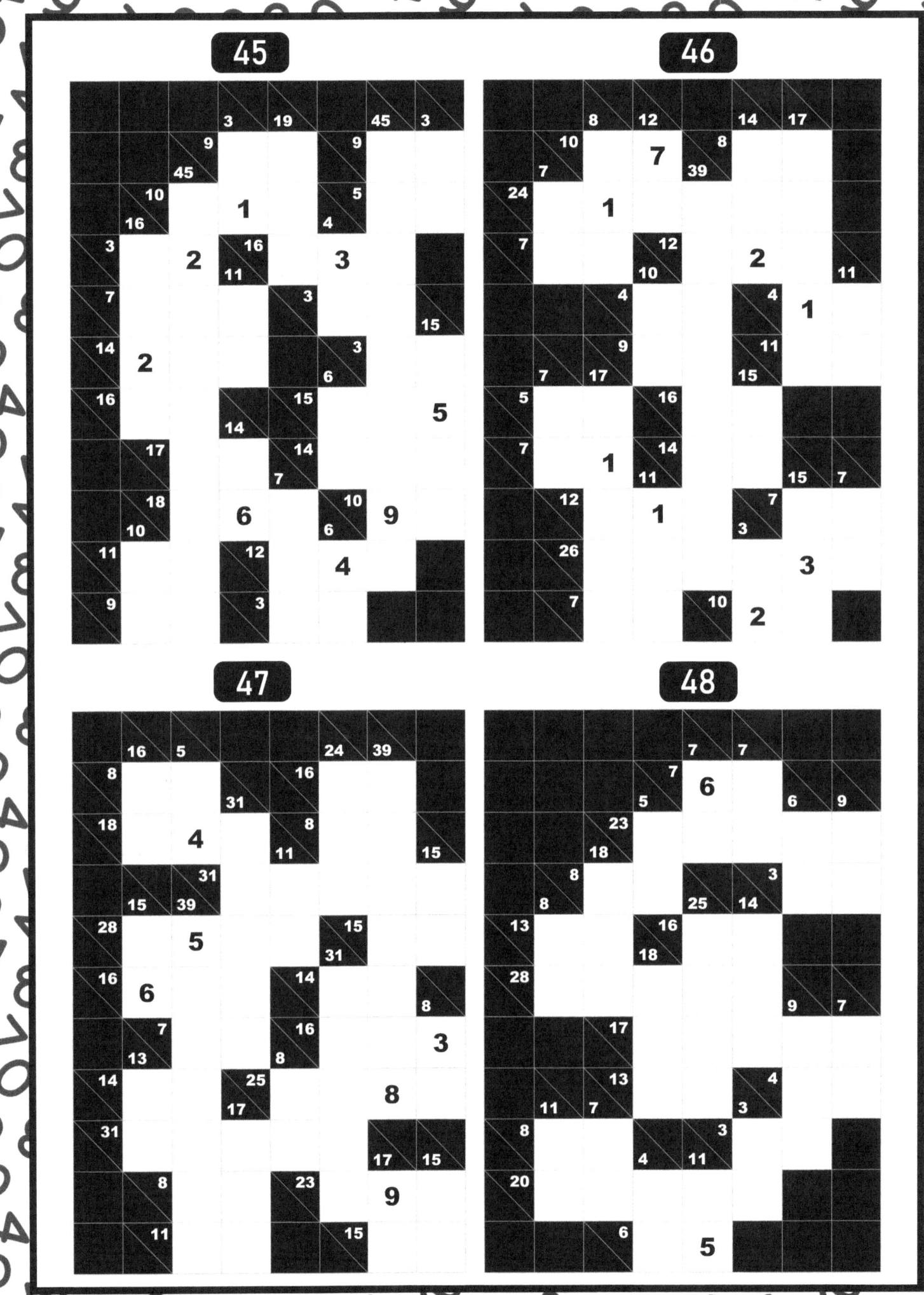

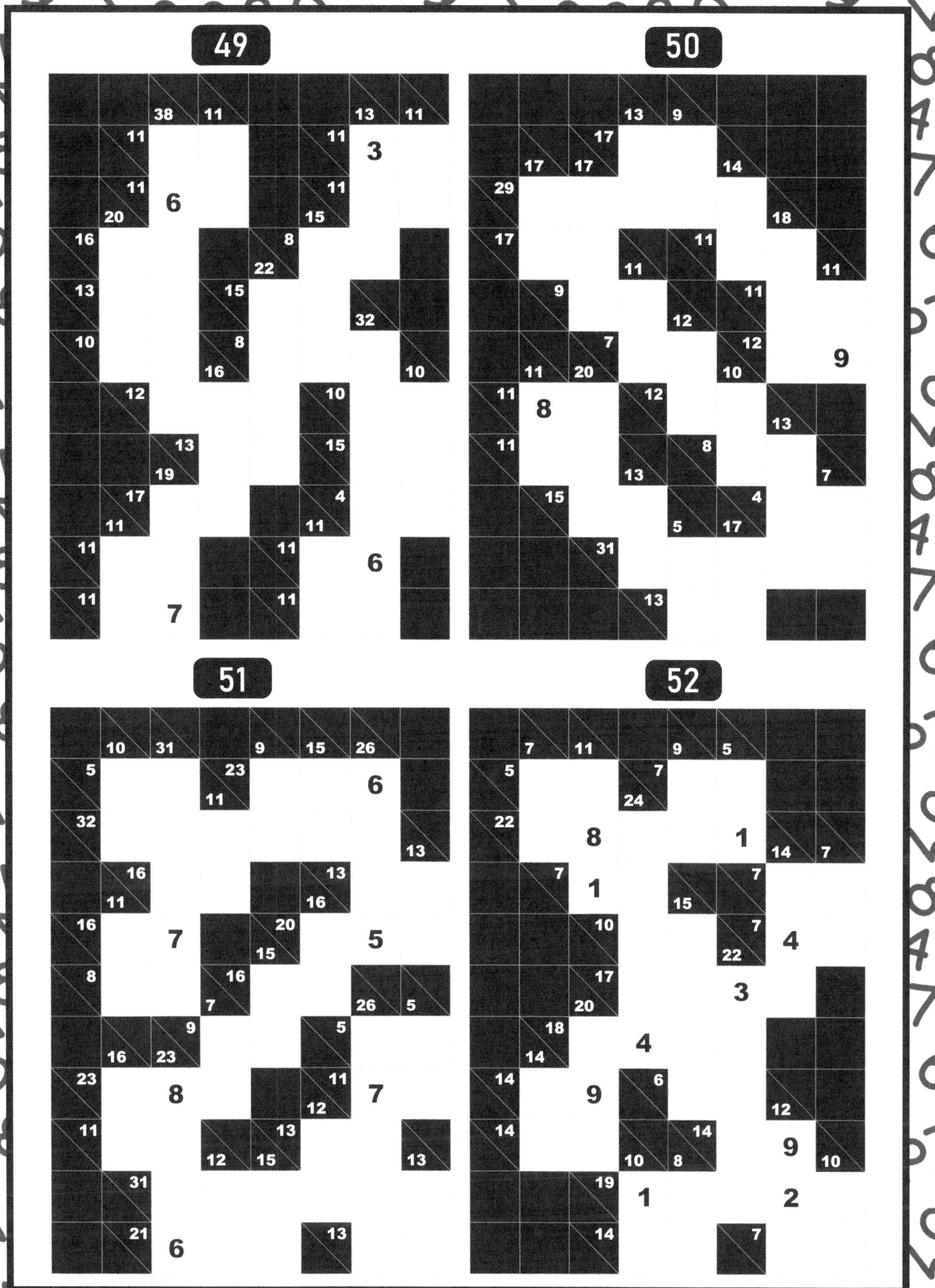

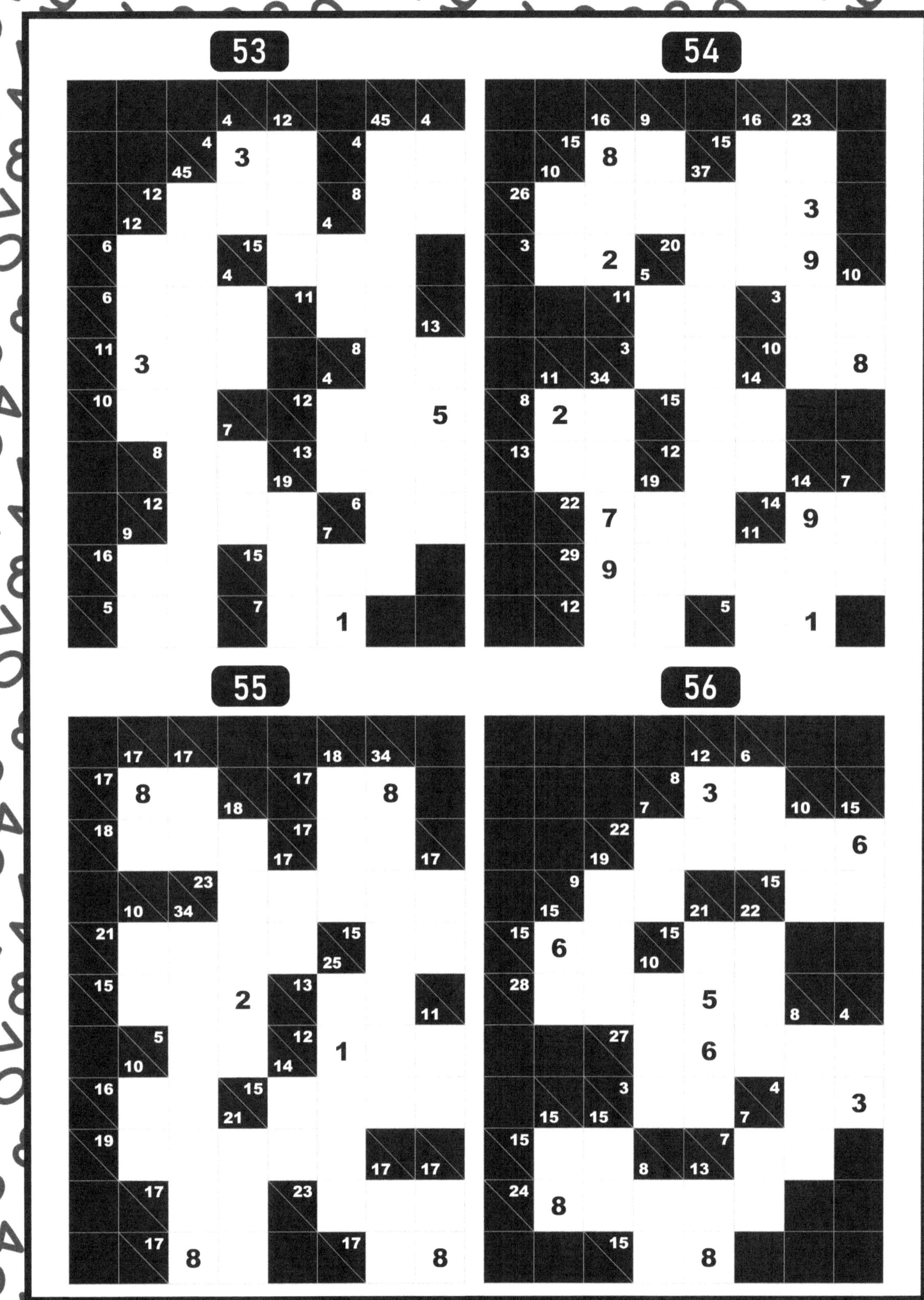

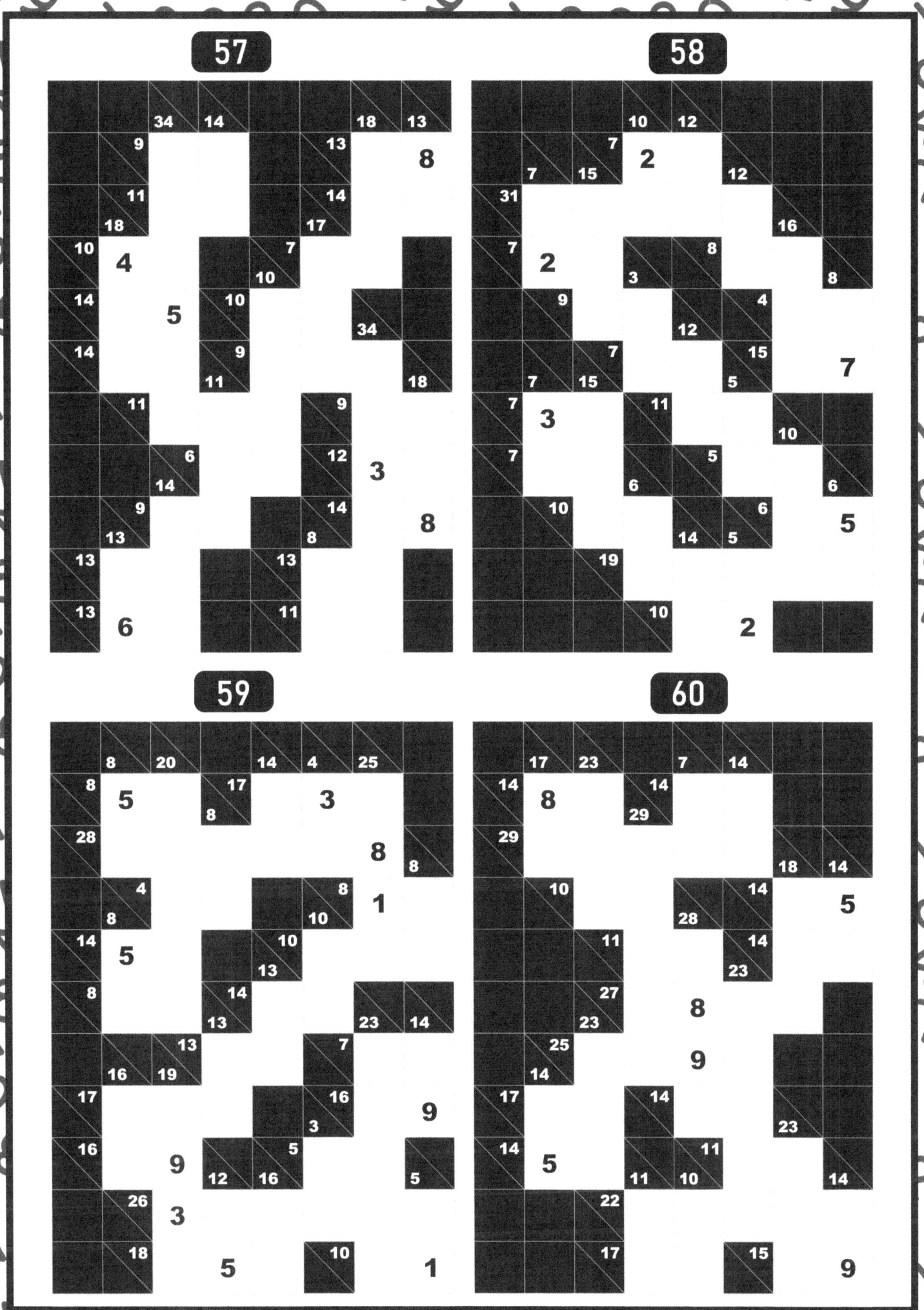

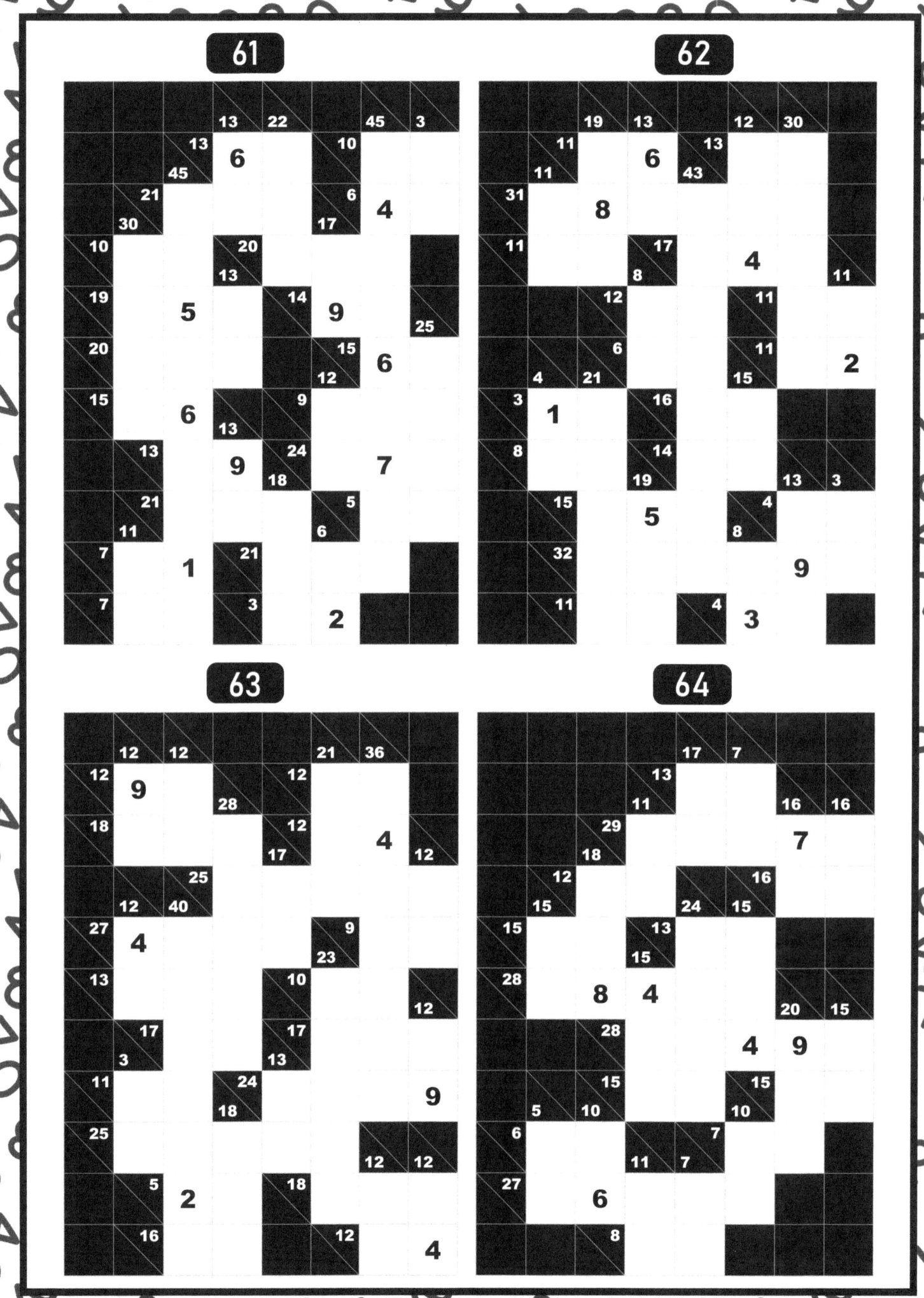

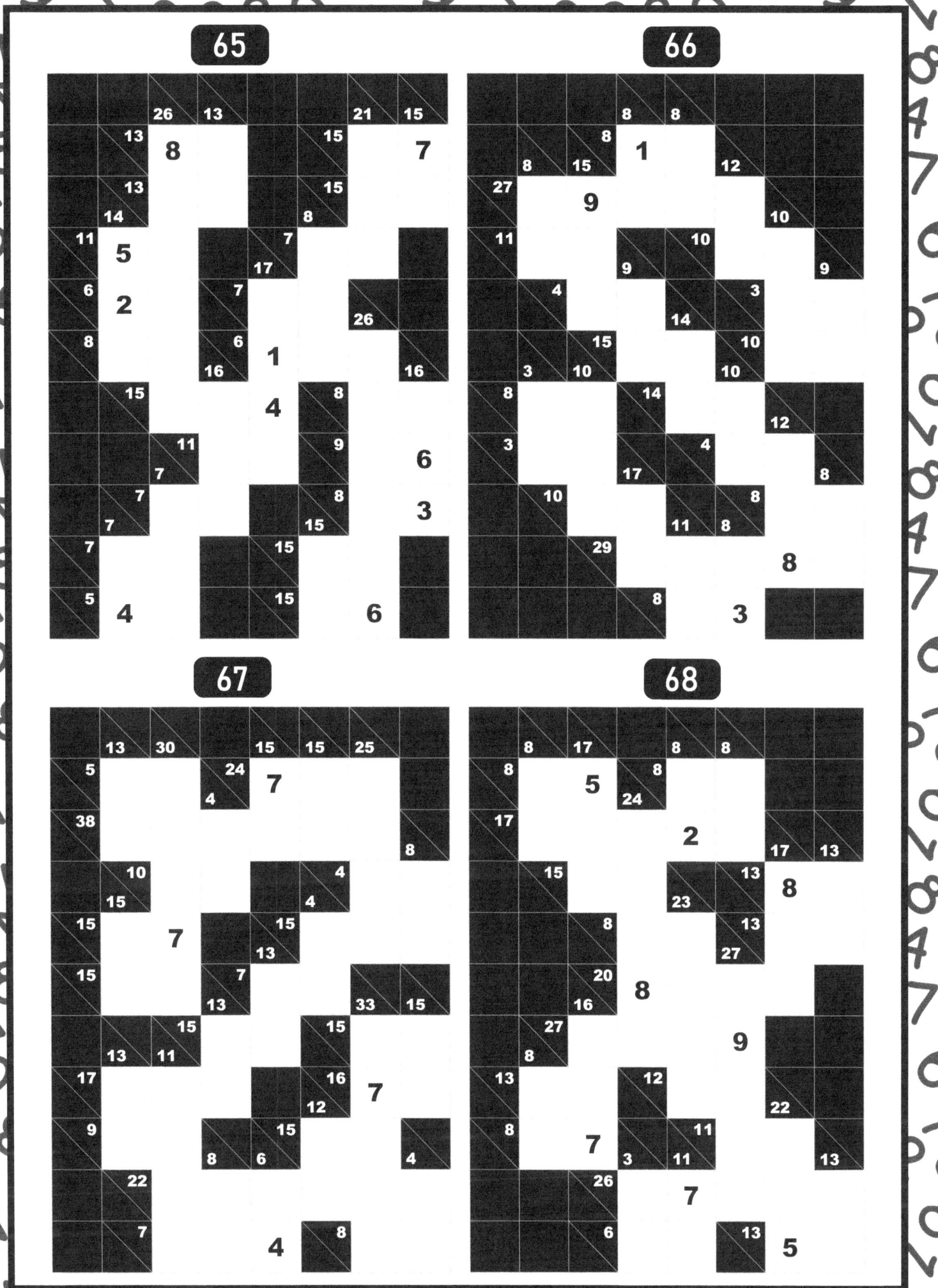

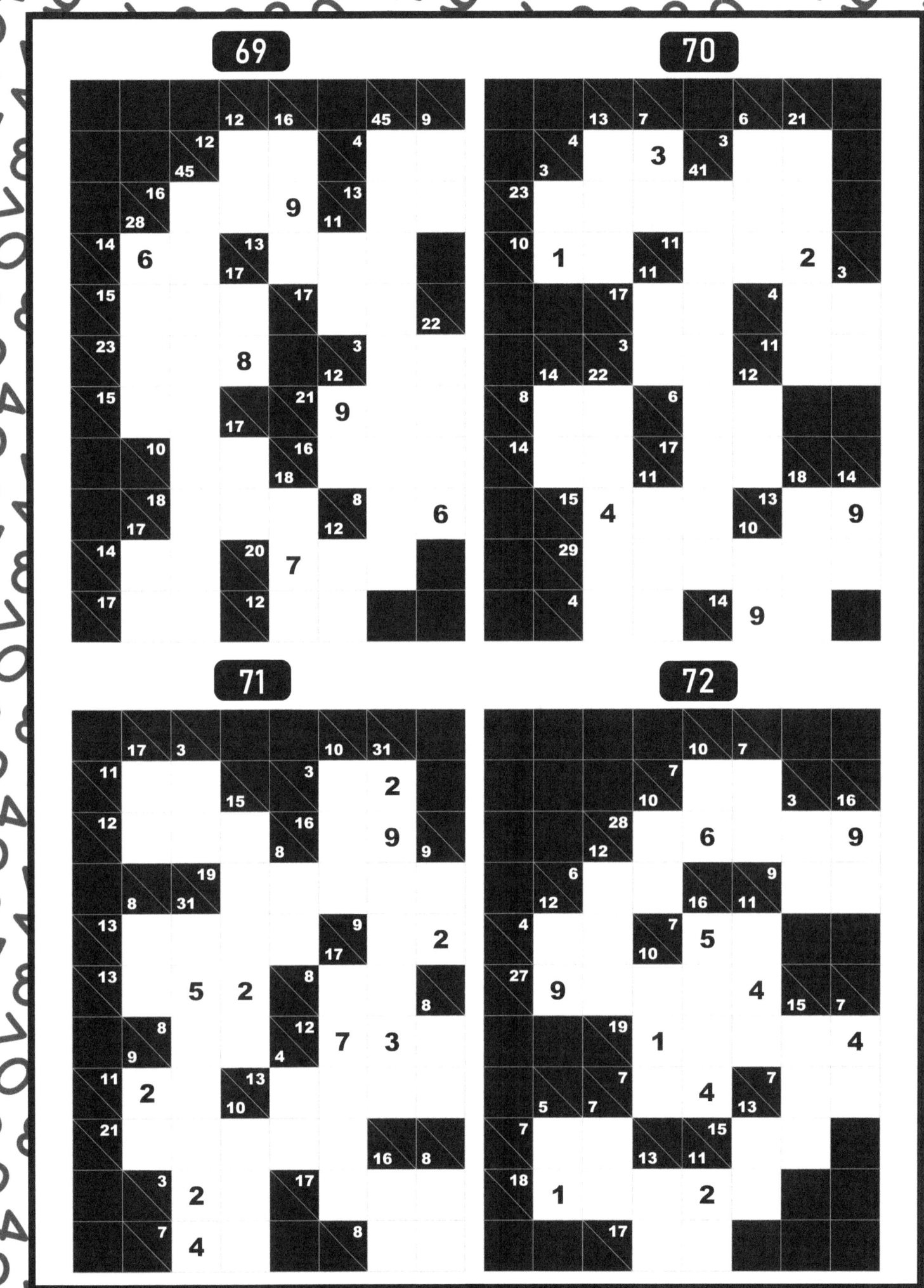

KAKURO

PUZZLE BOOK FOR ADULTS

-MEDIUM-

9X13

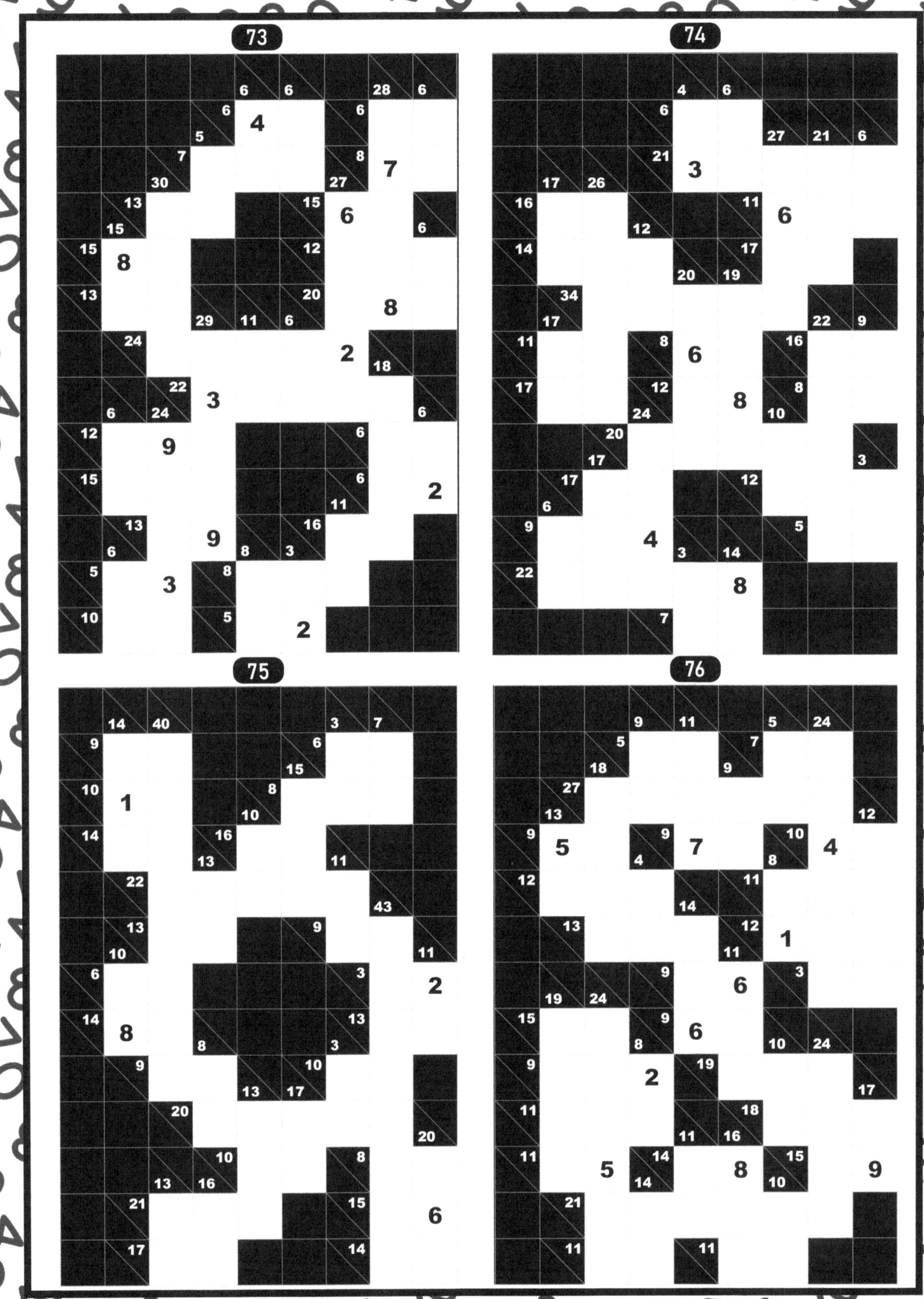

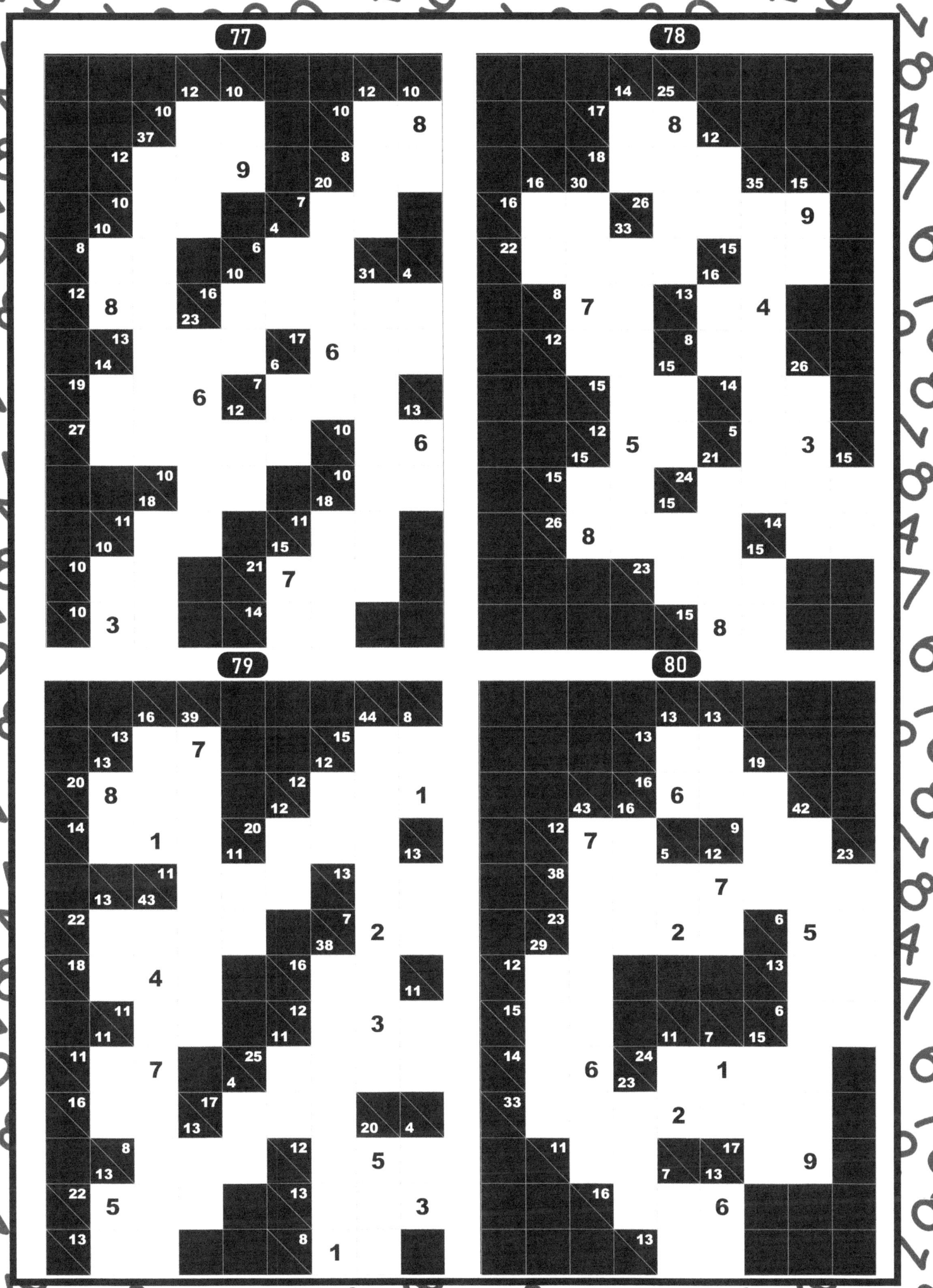

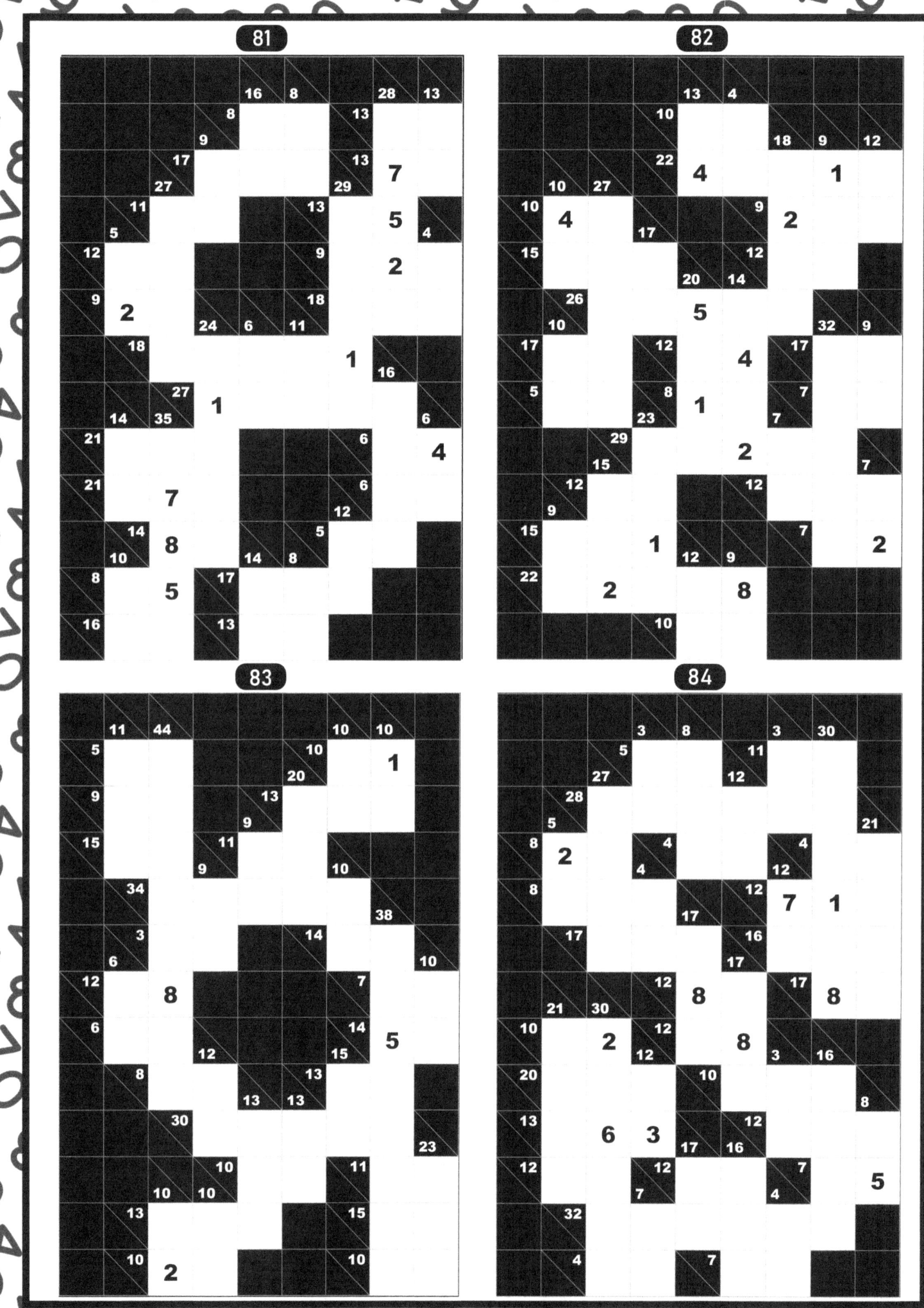

85

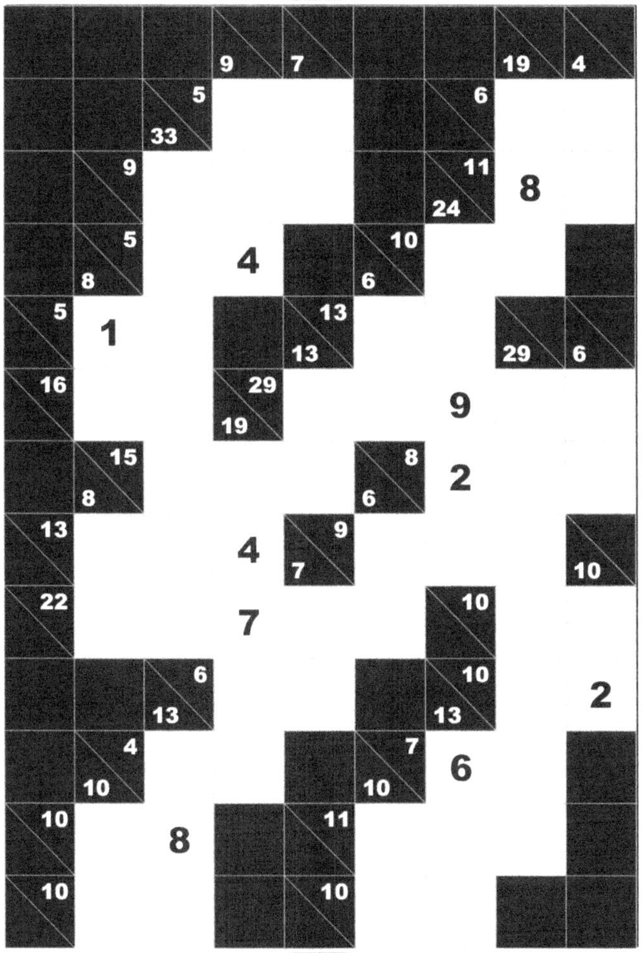
9 7 19 4
5 33 6
9 11 24 8
5 8 4 10 6
5 1 13 13 29 6
16 29 19 9
15 8 8 6 2
13 4 9 7 10
22 7 10
6 13 10 13 2
4 10 7 10 6
10 8 11
10 10

86

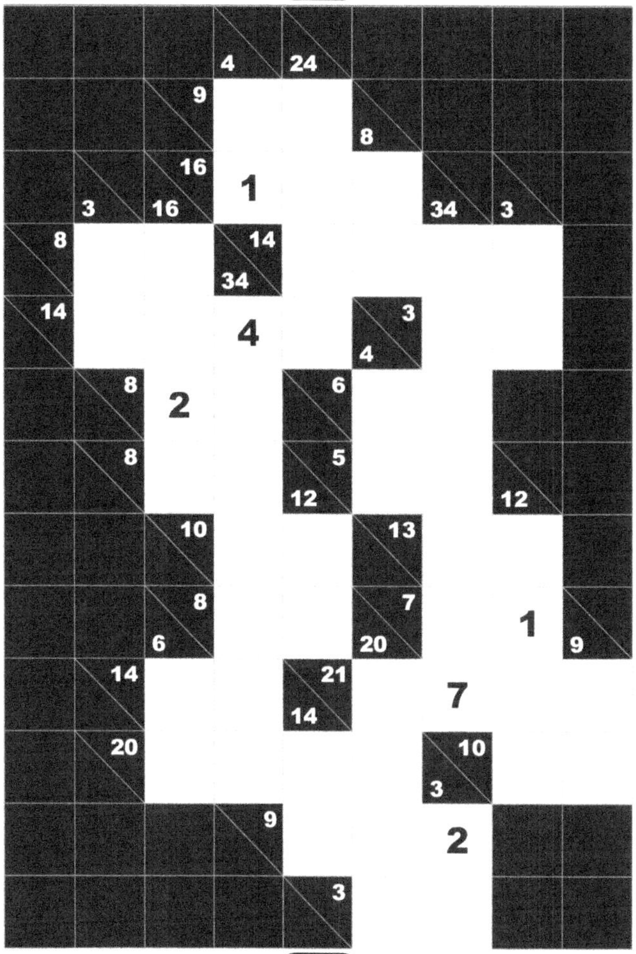
4 24
9 8
16 3 16 1 34 3
8 14 34
14 4 3 4
8 2 6
8 5 12 12
10 13
8 6 7 20 1 9
14 21 14 7
20 10 3
9 2
3

87

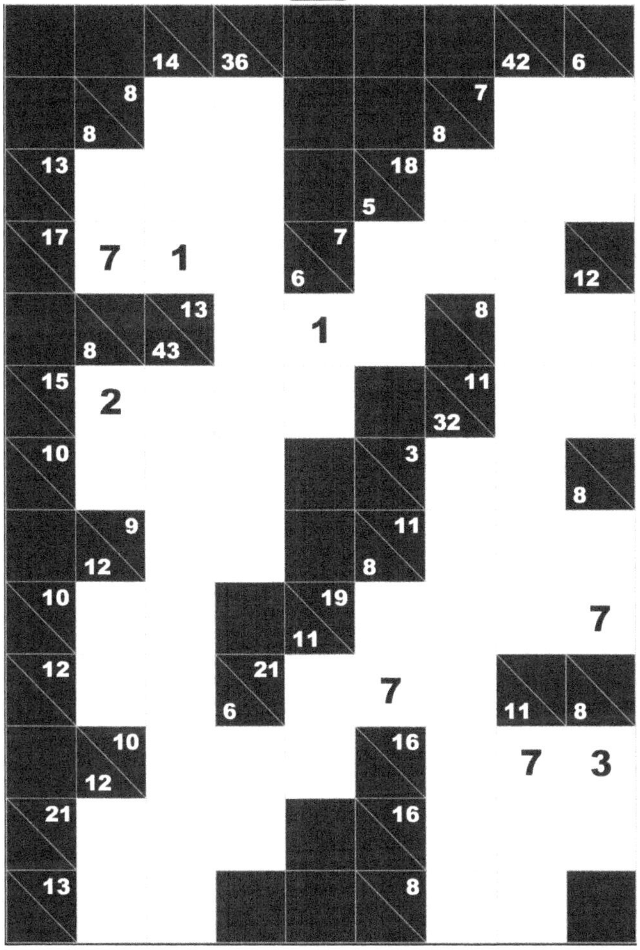
14 36 42 6
8 8 7 8
13 18 5
17 7 1 7 6 12
13 8 43 1 8
15 2 11 32
10 3 8
9 12 11 8
10 19 11 7
12 21 6 7 11 8
10 12 16 7 3
21 16
13 8

88

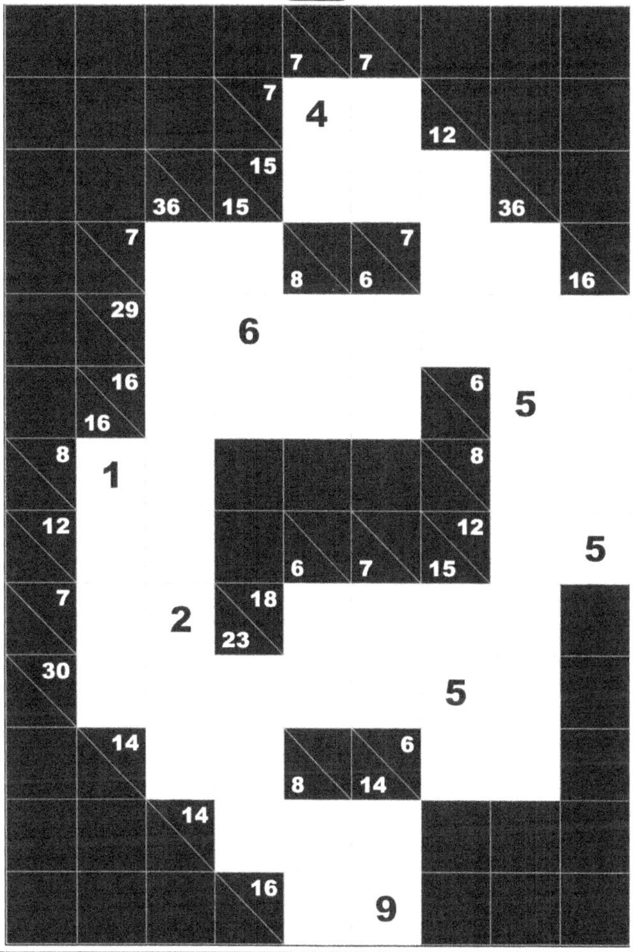
7 7
7 4 12
15 36 15 36
7 8 7 6 16
29 6
16 16 6 5
8 1 8
12 12 6 7 15 5
7 2 18 23
30 5
14 6 8 14
14
16 9

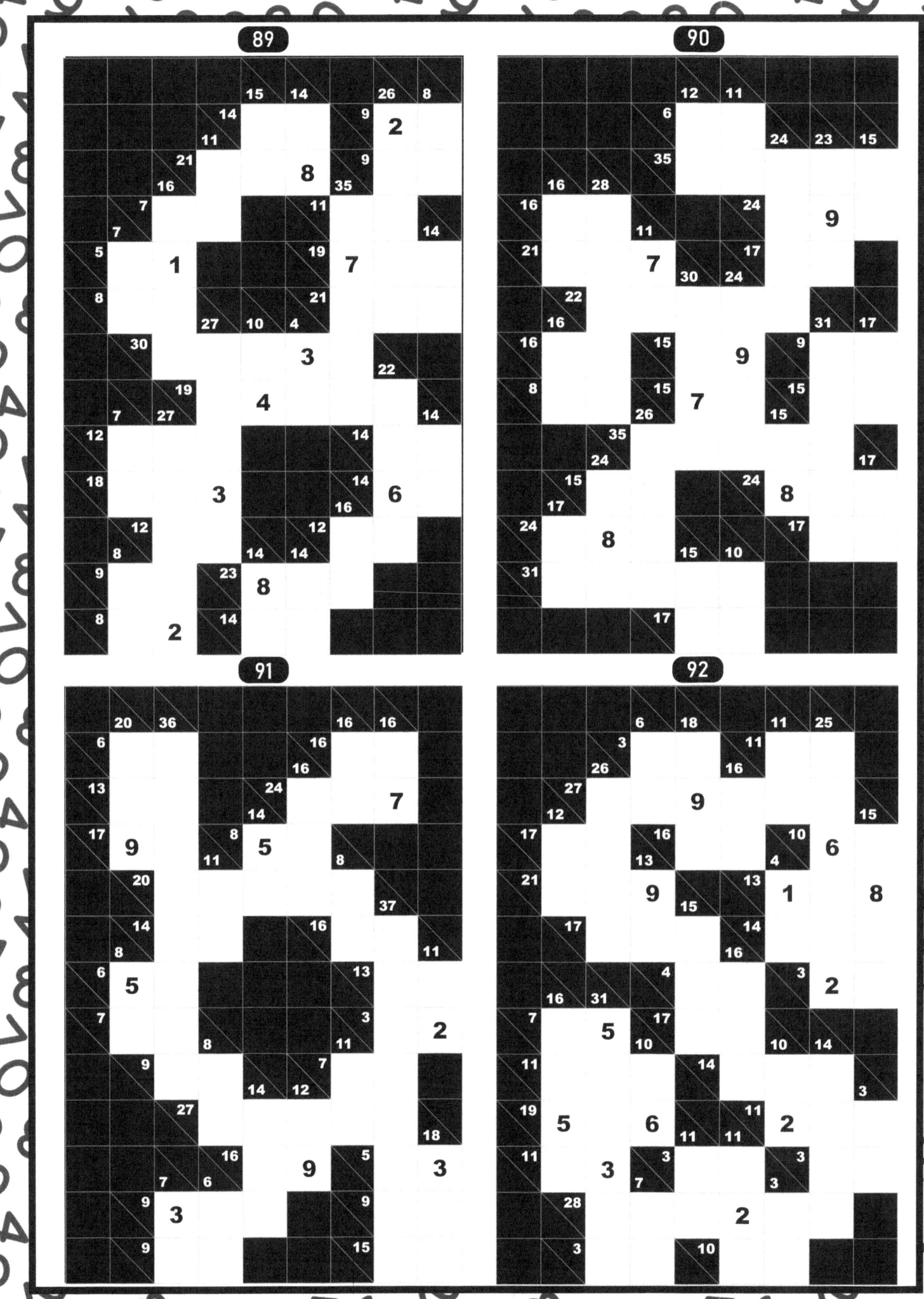

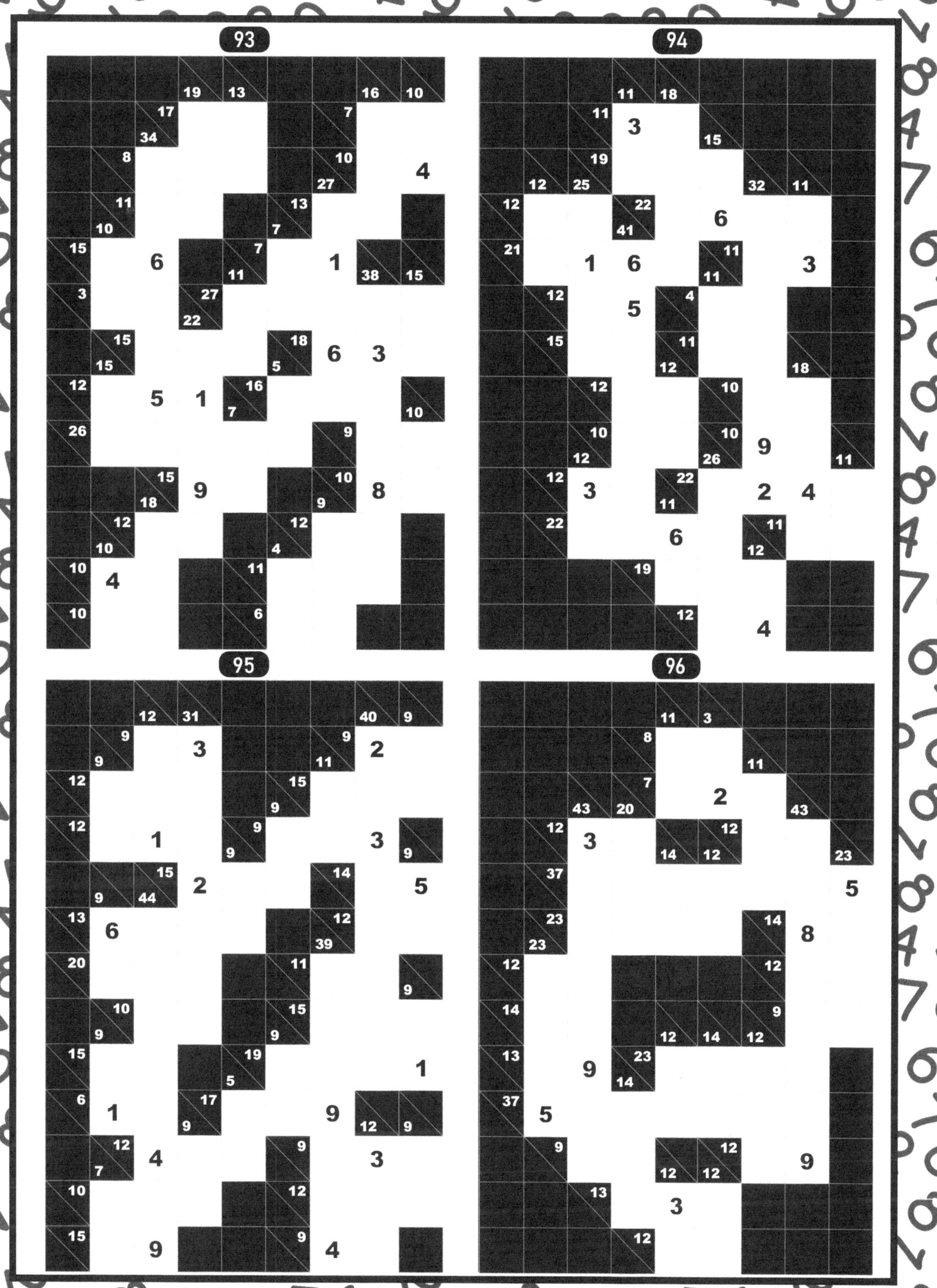

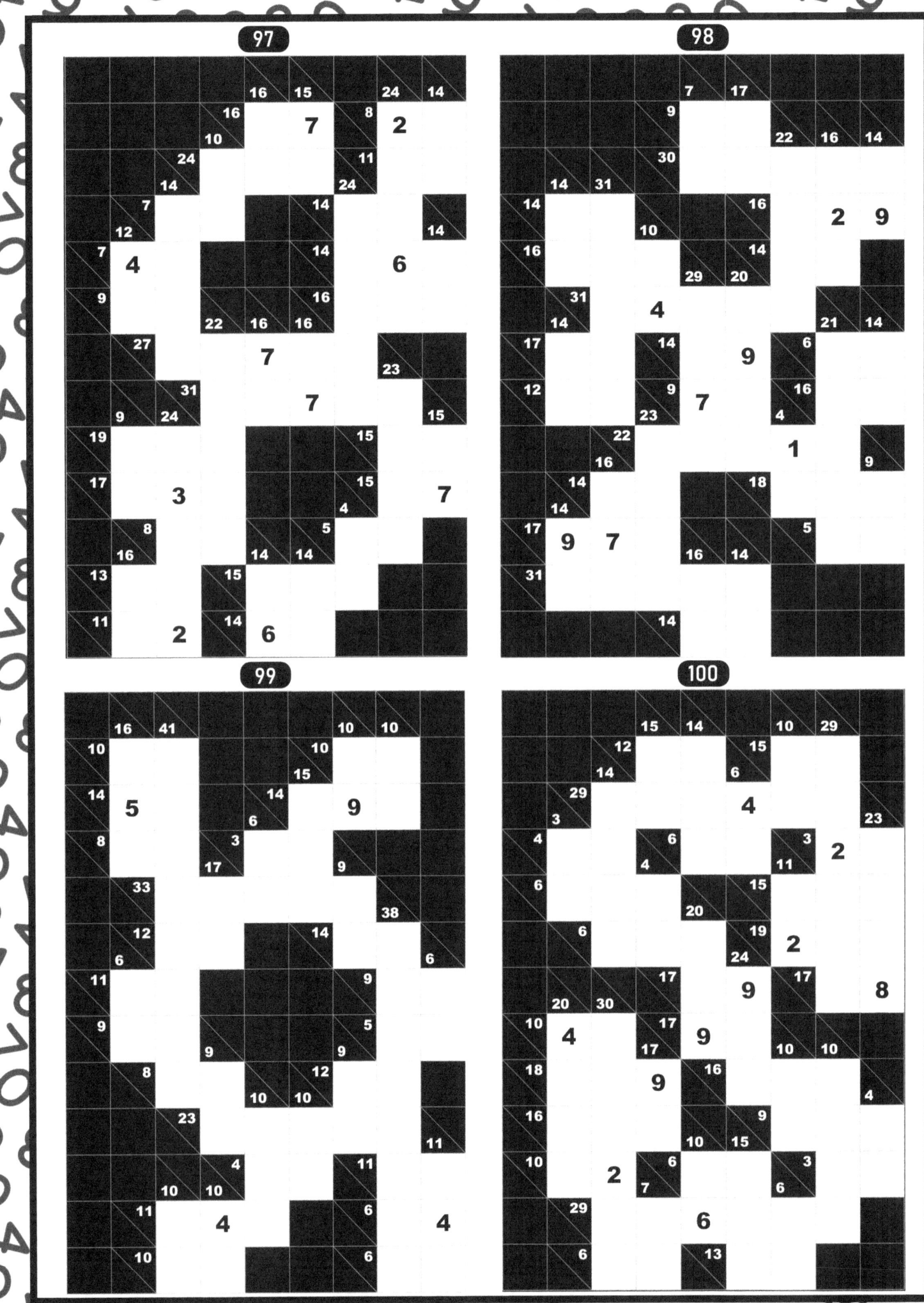

KAKURO

PUZZLE BOOK FOR ADULTS

-Solutions-

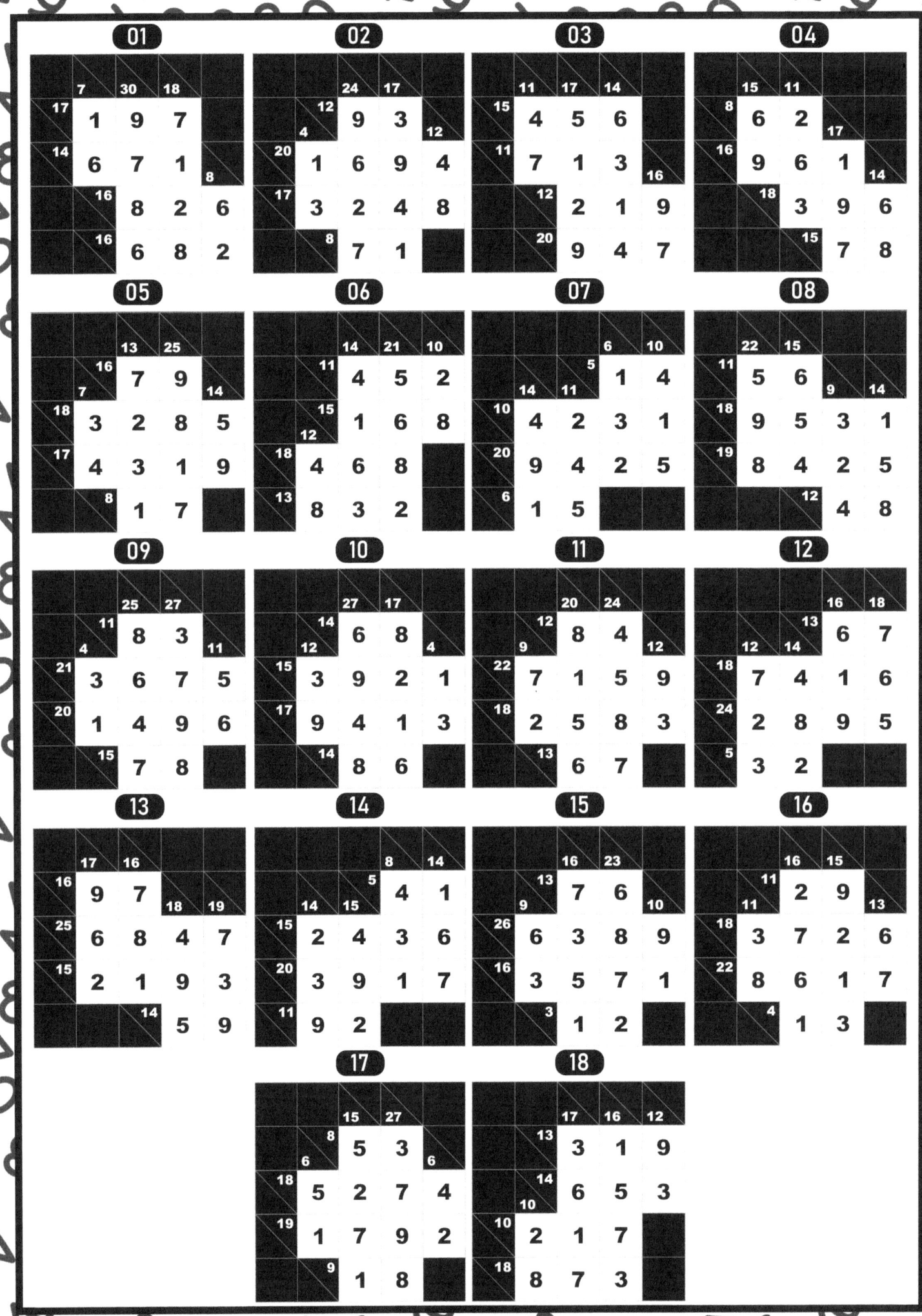

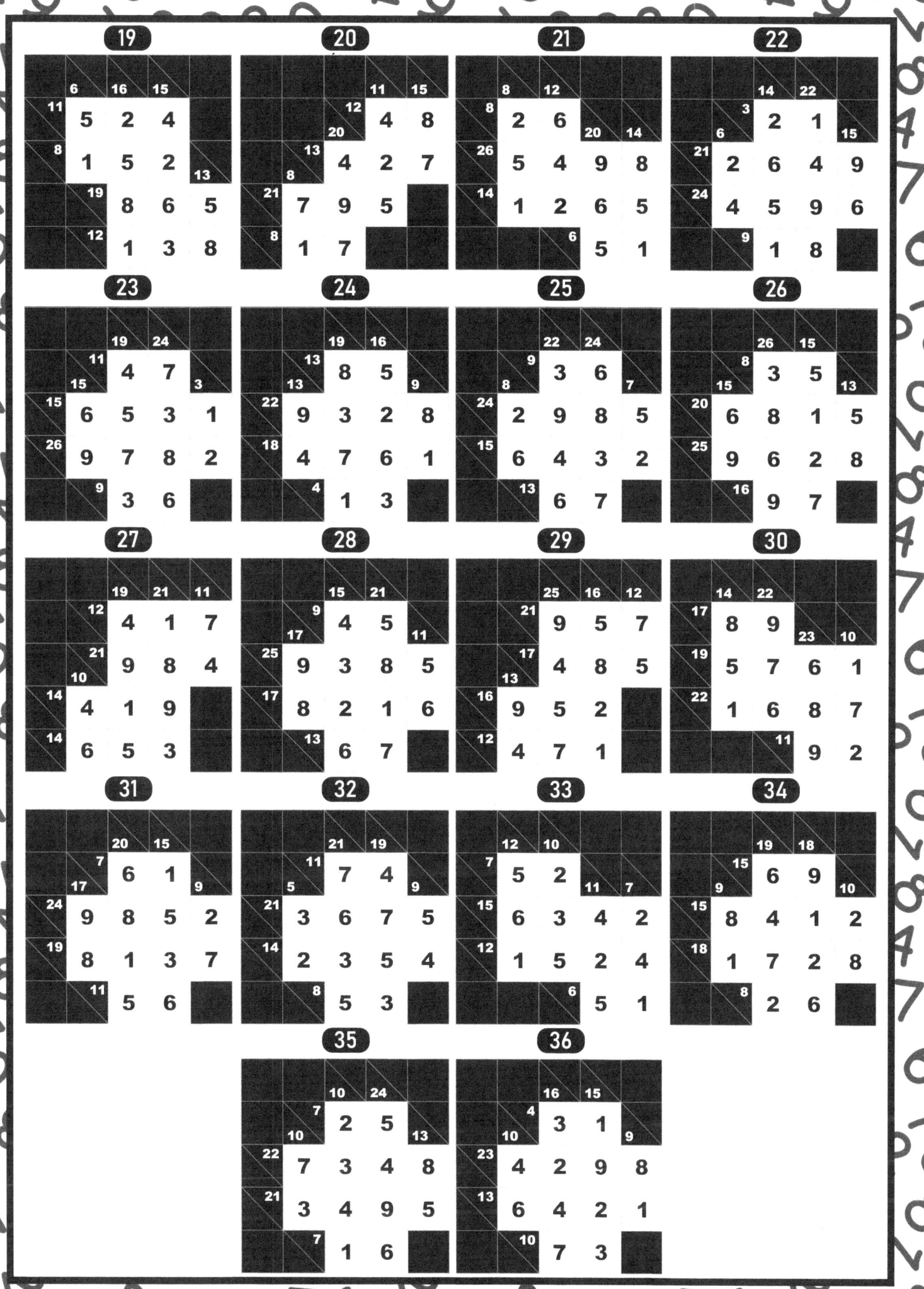

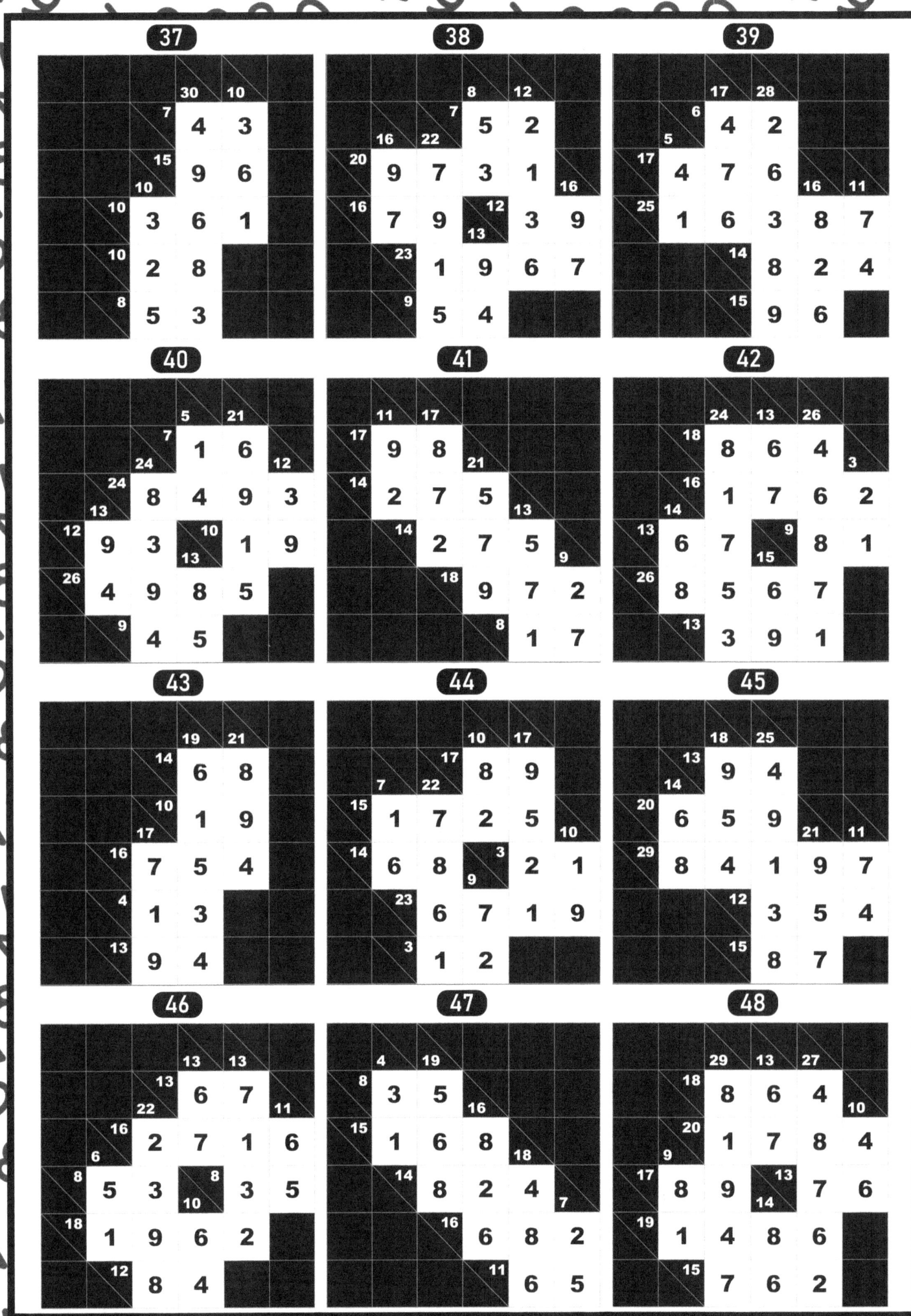

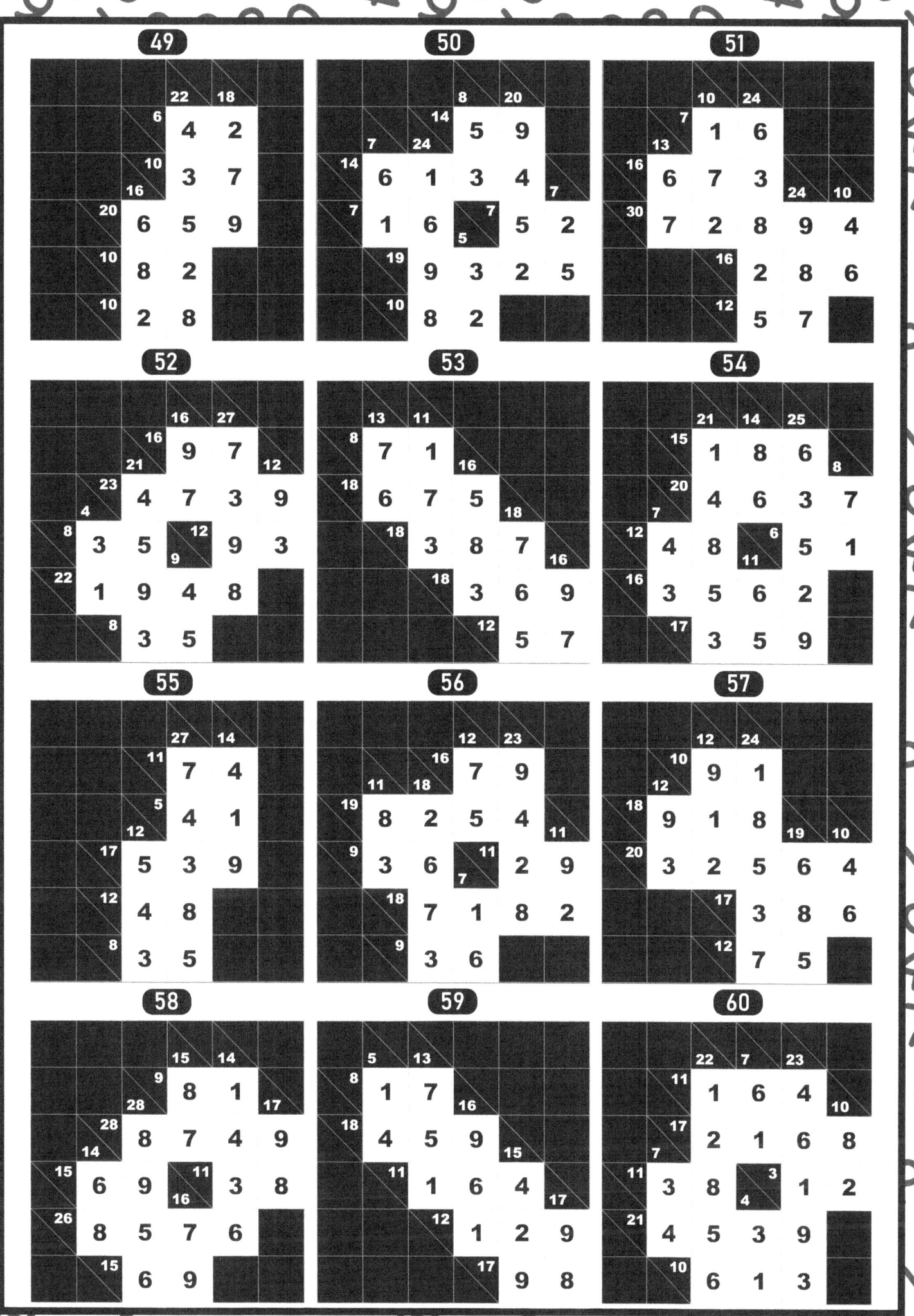

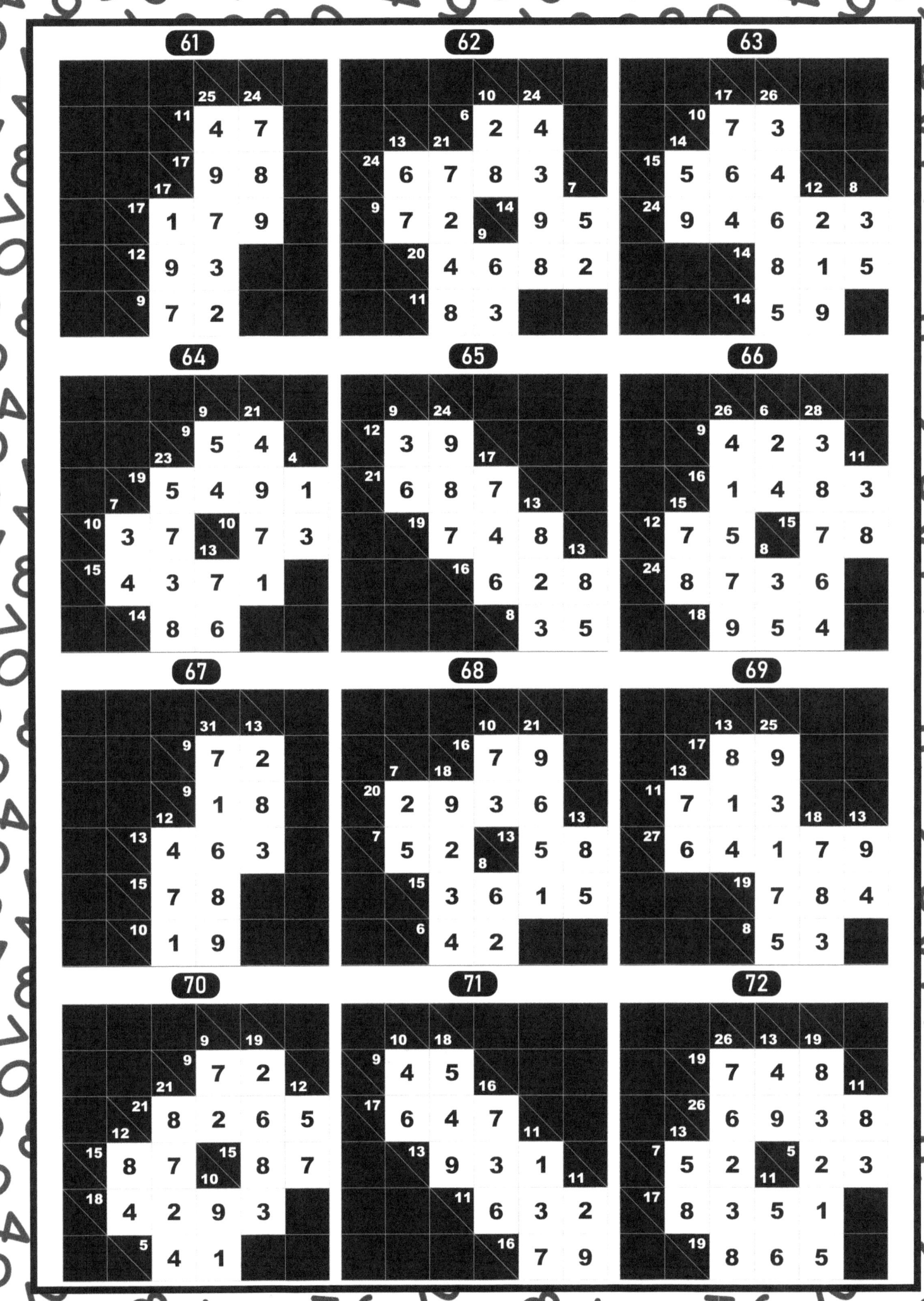

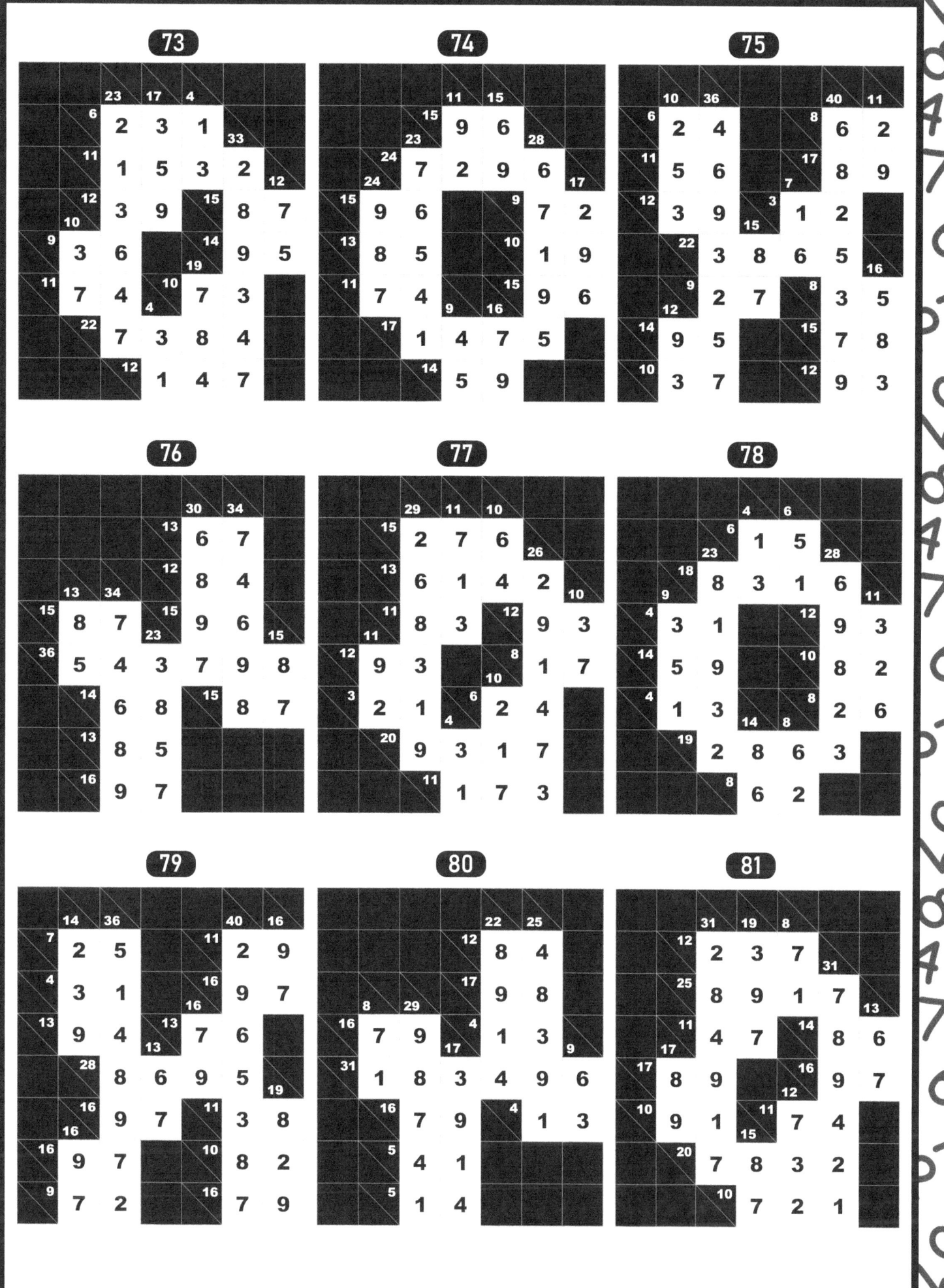

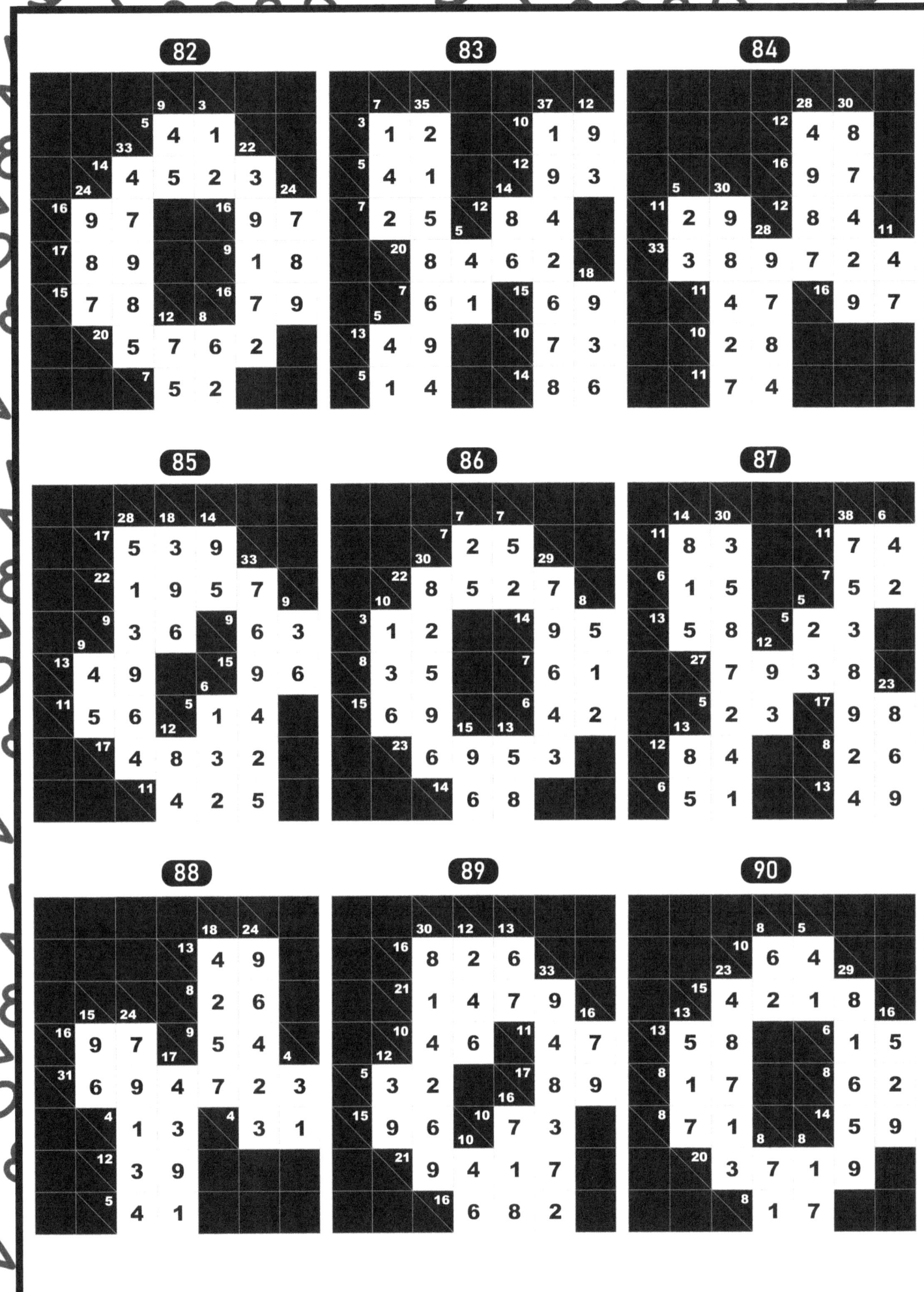

100

					24	29	
				16	9	7	
			14	6	8	
	12	29	17	8	9	8
12	5	7	23			
26	7	5	6	1	4	3
	9	2	7	6	1	5
	11	9	2			
	14	6	8			

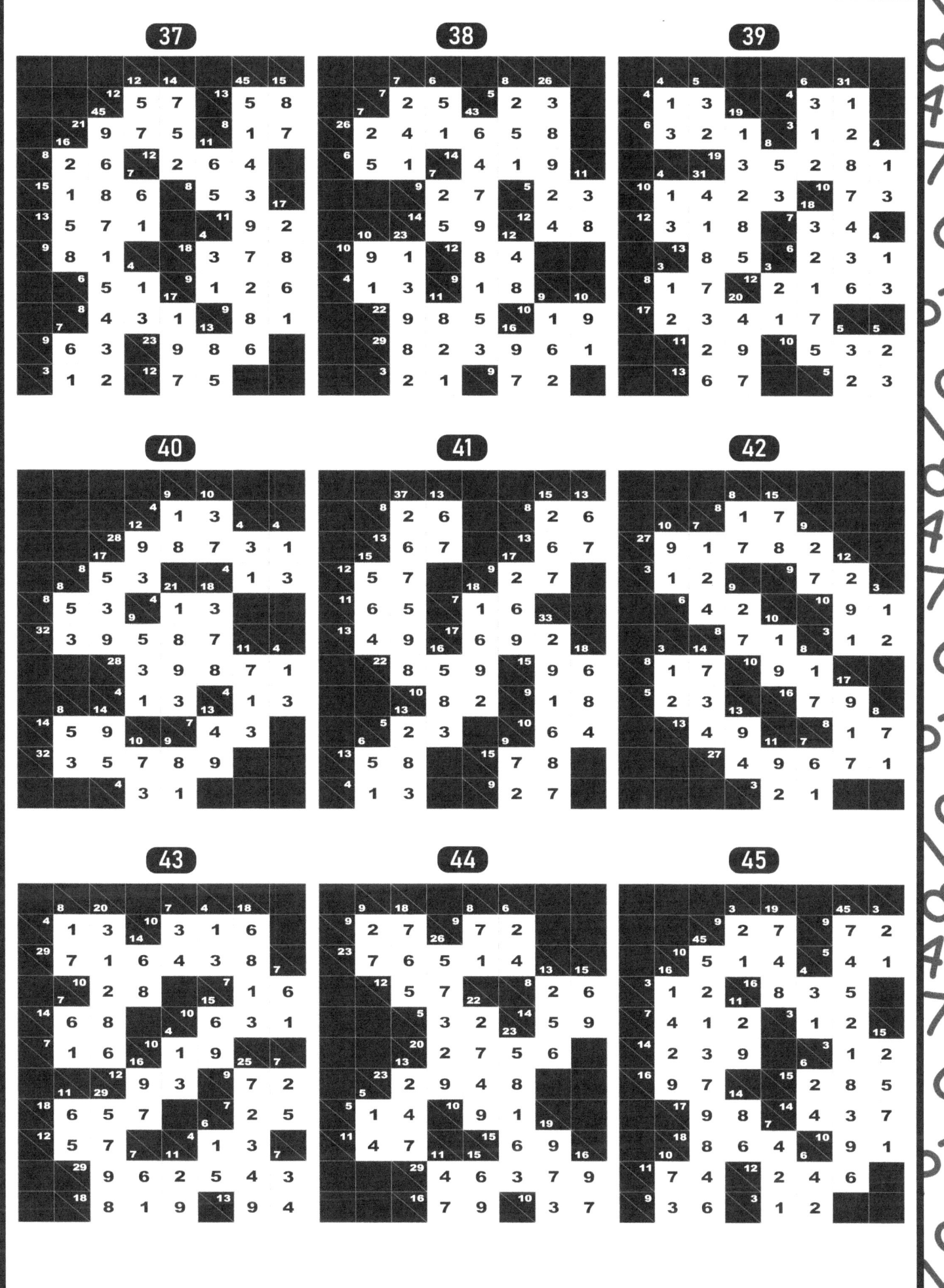

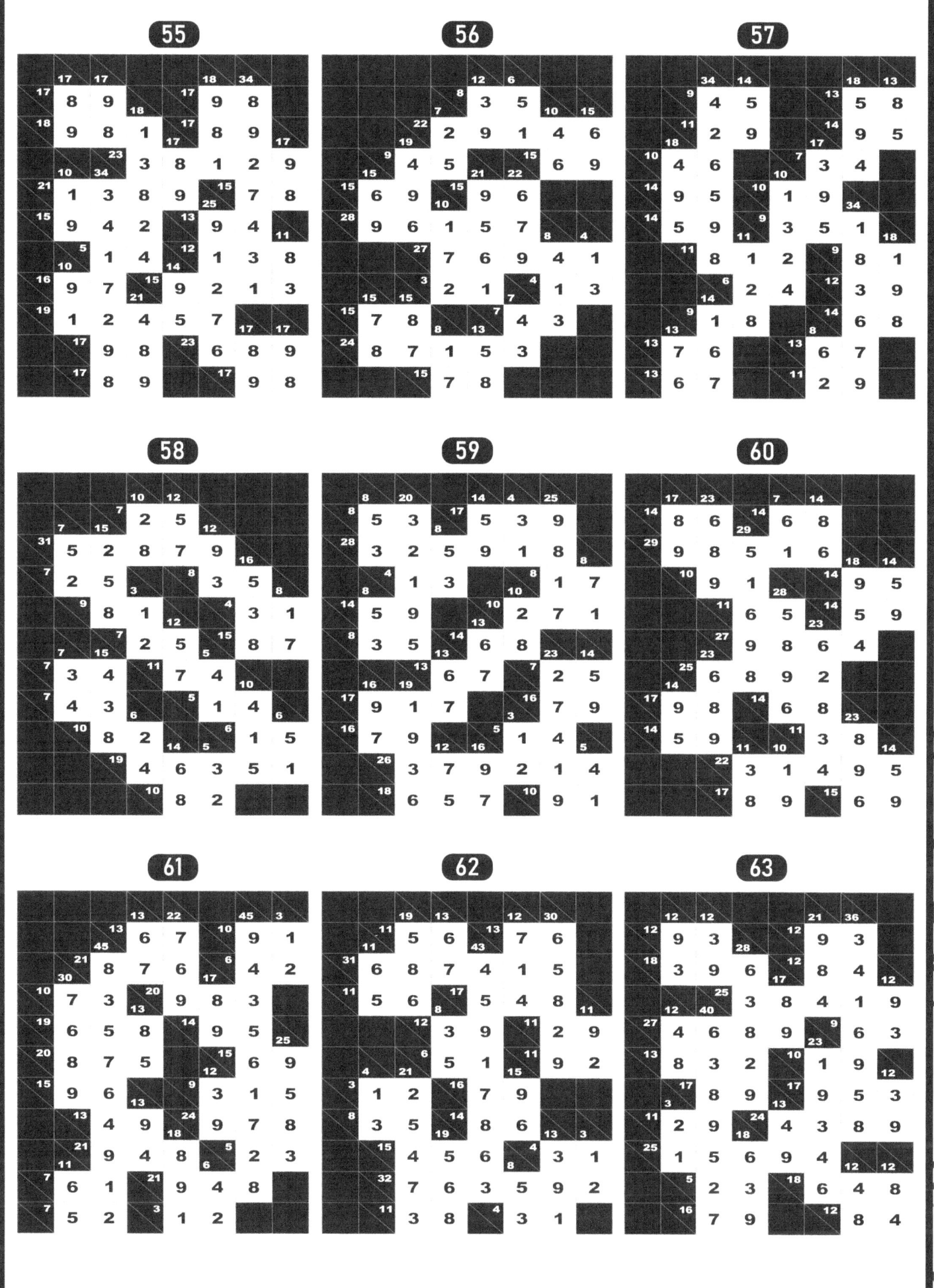

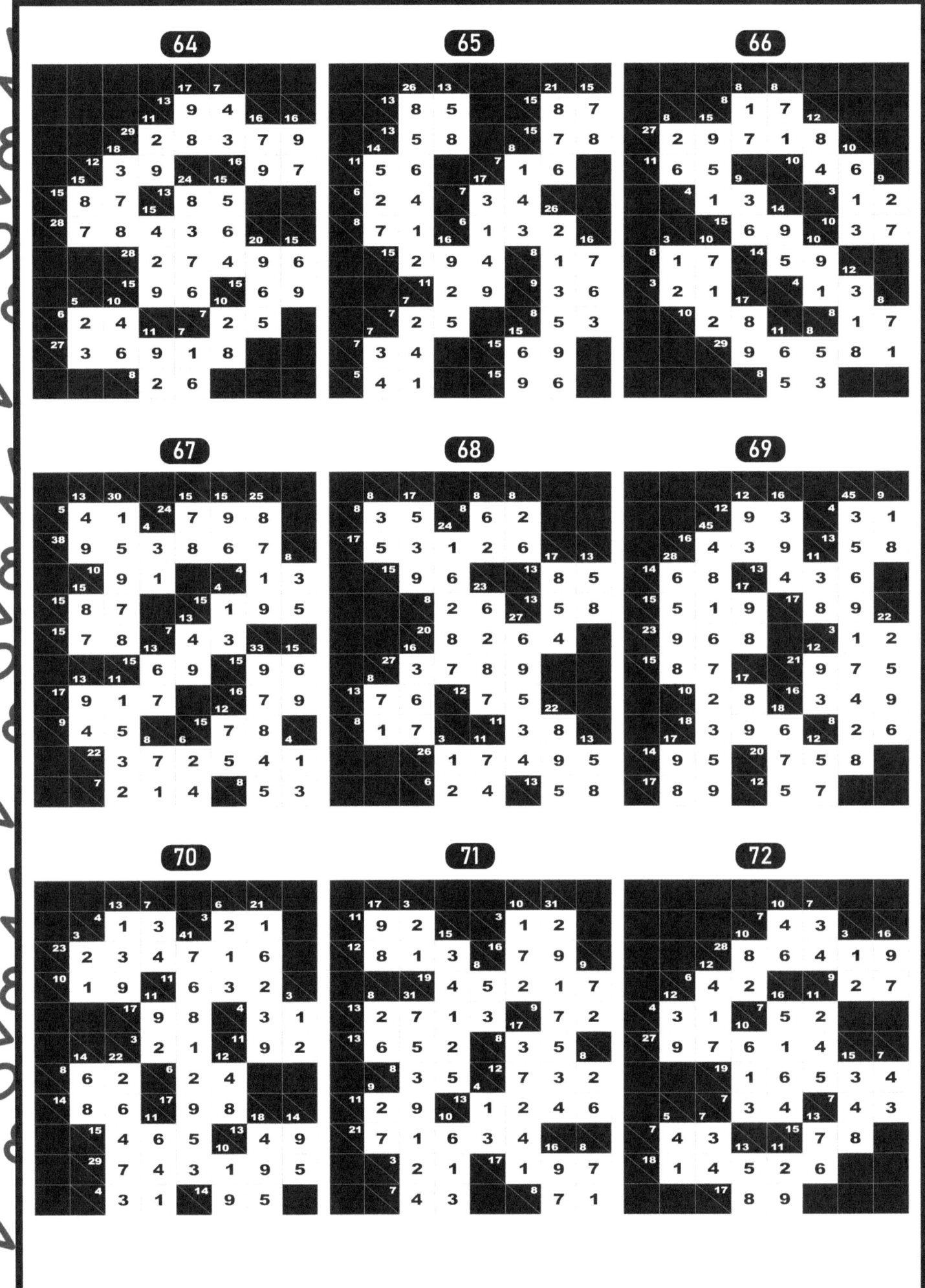

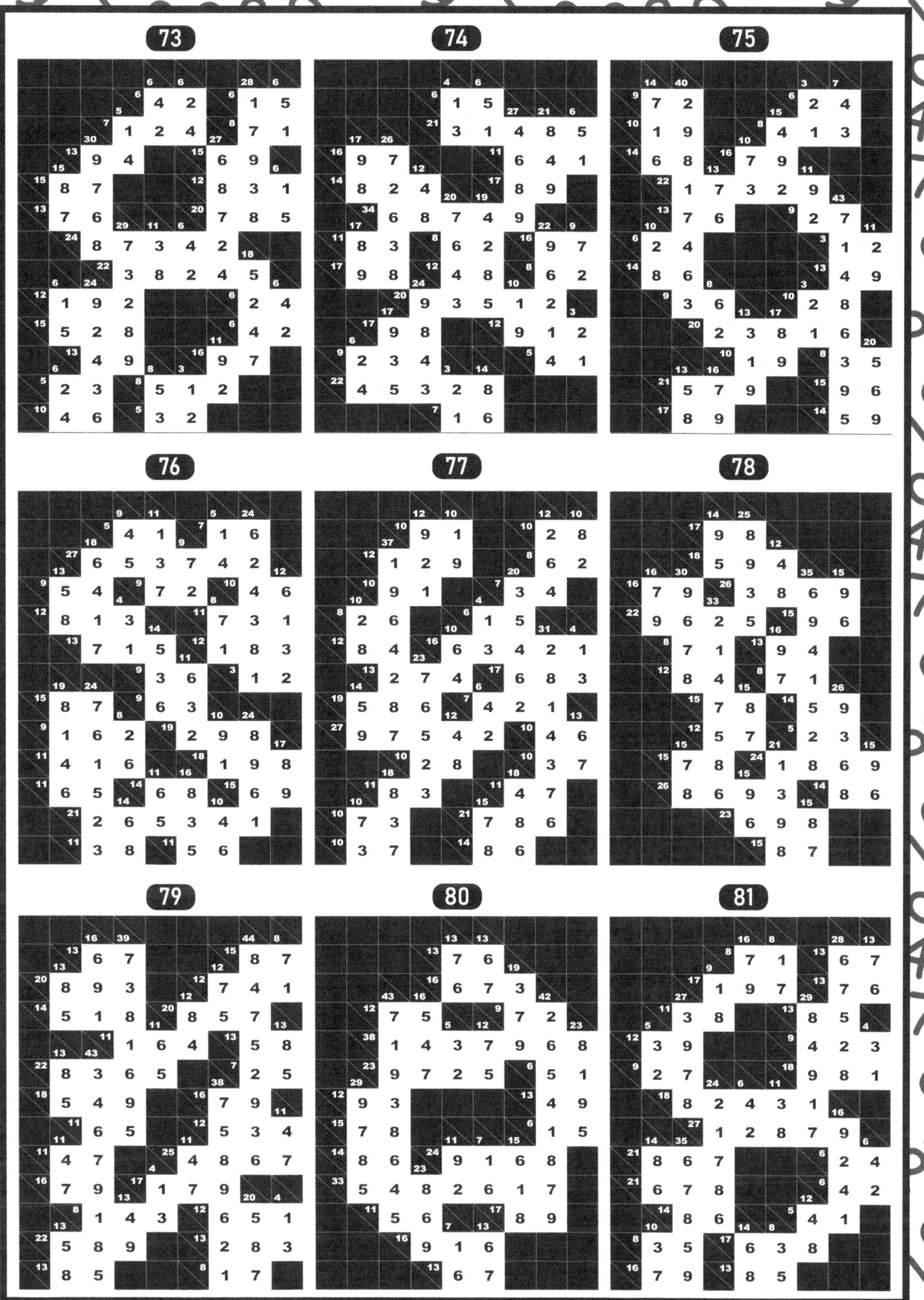

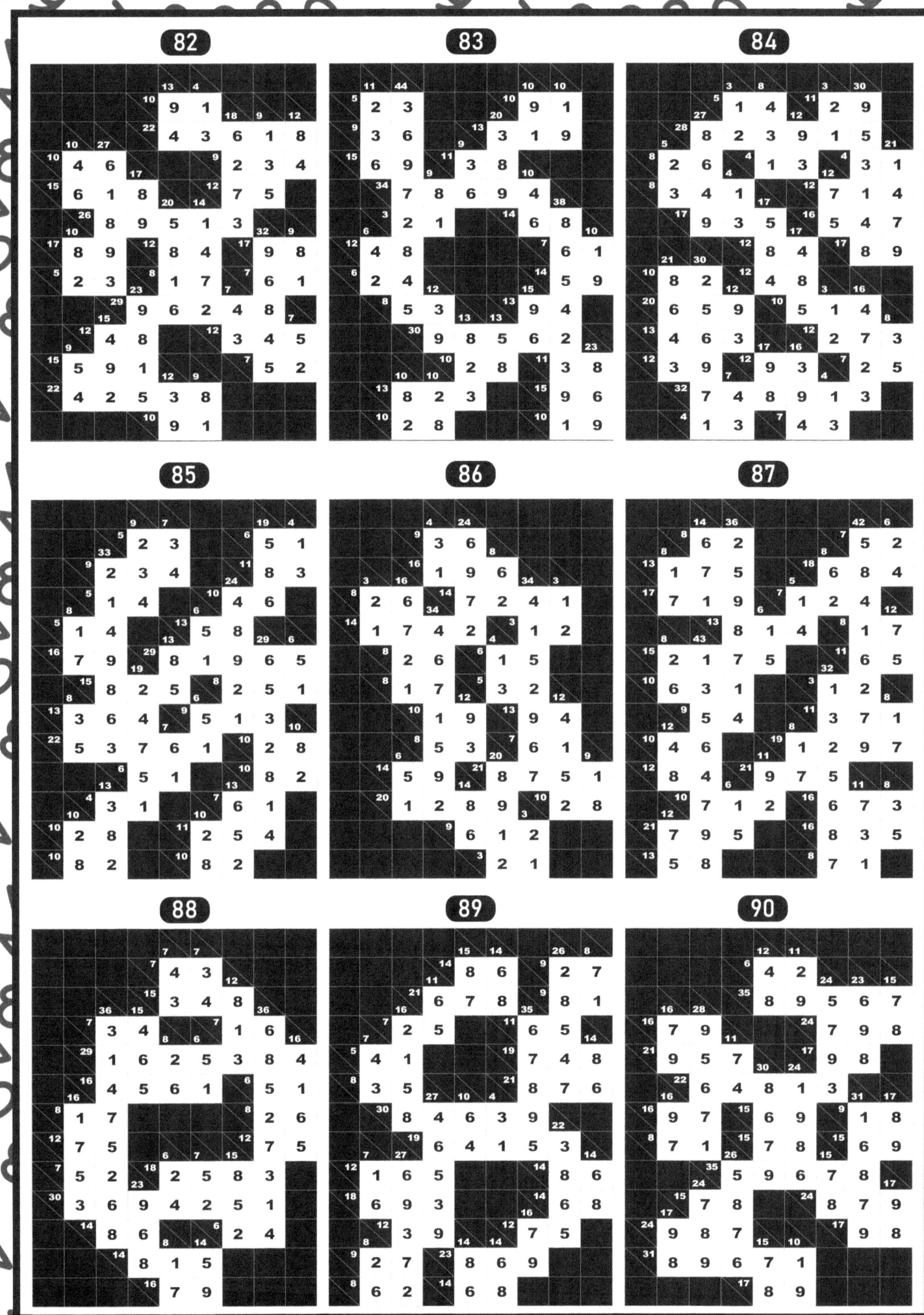

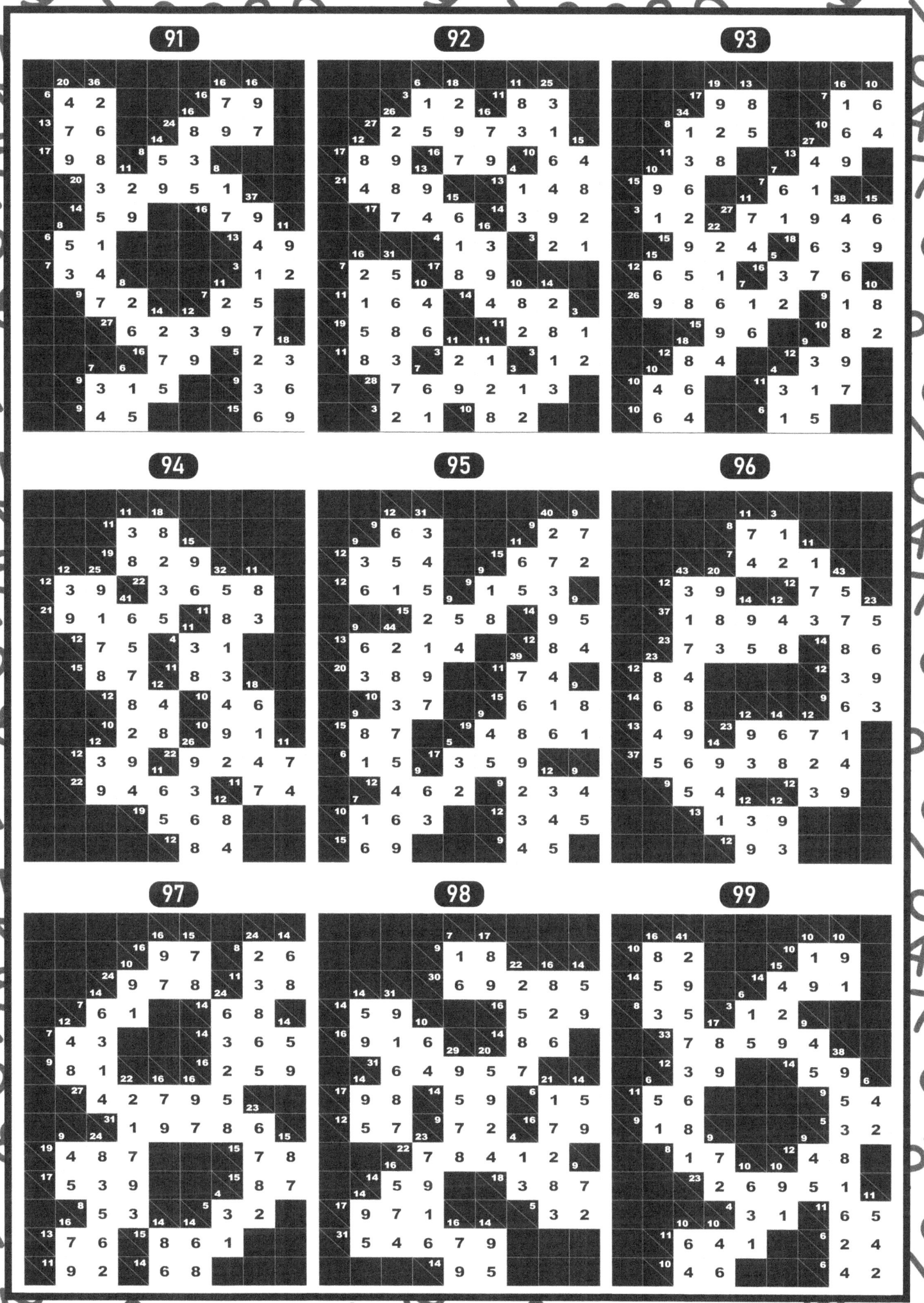

100

			15	14		10	29	
		12	9	3	15/6	9	6	
	29/14	8	6	7	4	1	3	23
4/3	1	3	6/4	4	2	3/11	2	1
6	2	1	3	15/20	15	9	1	5
	6	2	1	3	19/24	2	8	9
20	30		17	8	9	17	9	8
10	4	6	17/17	9	8	10	10	
18	1	8	9	16	7	8	1	4
16	7	1	8	9/10	15	2	4	3
10	8	2	6/7	4	2	3/6	2	1
	29	9	5	6	4	2	3	
	6	4	2	13	9	4		